MATRIX MANAGEMENT

Matrix management

Edited by
KENNETH KNIGHT

Gower Press Special Study

Published by

Gower Press, Teakfield Limited,
Westmead, Farnborough, Hants., England

ISBN 0 566 02076 9

Printed in Great Britain by Biddles Ltd, Guildford, Surrey

Contents

Preface

This book has been written and edited with an eye on two audiences, but more on one than the other. While I hope that teachers, researchers and students of organisation and management will find it a useful source of information, reference material and even some ideas, our main concern has been for those who face at first hand the problems of structuring, managing and working in complex organisations. The book is not intended as a contribution to the sociology of organisations, but as an attempt to document and formulate an area of growing practical concern to many managers.

The book had its origin in a series of exploratory seminars and workshops on matrix organisation held at Brunel University over the last two years during which we had the chance to discuss the concepts and problems of matrix management with about eighty managers and specialists from a variety of organisations, many of them already attempting to operate some form of matrix. Most of the contributors to the book also took part in these seminars, and I am sure they were stimulated as much as I was by these discussions to look more closely at the implications of matrix management.

In my role as editor I have deliberately abstained from any attempt to impose a common set of assumptions or single frame of reference on all the contributions: my aim has been coherence rather than consistency. I believe this has resulted in a wider range of experiences, points of view and differences of emphasis being made available to the reader, and that it has produced a fuller and more realistic reflection of the field we set out to explore. Some might disagree with this approach; my own view is that, in dealing with as complex and diverse a field as that of human organisation, it is a positive advantage to be presented with a choice of viewpoints as no single one is likely to be wholly adequate.

The individual authors' names appear at the head of their respective chapters. The chapters which do not carry a name were written by myself.

Acknowledgements

I would like to thank all the participants in our matrix seminars who helped us to come to grips with the problems, and all those who, in our recent survey and on other occasions, took time and trouble to provide information about their organisations. My thanks are especially due to Bryan Reader who went to considerable pains to provide information on two special areas of matrix management, to Ken Atterbury, Dr J. Basterfield, Arthur Johnston, Dr C. Nicholls, David Paskins, John Searle and Donald Videlo for enabling me to gain some detailed insights into specific organisations; to Professor J. B. Wilkinson for giving me the opportunity to share the experiences of a group of European and British R and D managers during a seminar of the European Industrial Research Management Association; to Dr R. Greenwood for his comments on the section on local government organisation and to Harold Bridger for giving me the benefit of his long experience in conducting consultancy and training for matrix and other complex organisations. I also want to express my gratitude to my colleagues in the Brunel Institute of Organisation and Social Studies for the stimulation they have provided in many internal discussions on aspects of organisation. The very considerable debt which I owe to the work of Elliott Jaques in clarifying the nature and structure of bureaucratic organisation will be apparent from many pages of the text; this does not mean, however, that he bears any responsibility for the views put forward in this book. The same applies to other colleagues, particularly Ralph Rowbottom, David Billis, Anthea Hey and Gillian Stamp, whose comments on parts of the manuscript helped me to review my approach to a number of issues. Finally, I want to thank Angela Simms for her essential help in typing and preparing the manuscript.

Brunel University   Kenneth Knight
1977

The contributors

Maureen Dixon is Assistant Professor in the Department of Health Administration, University of Toronto. From 1967 to 1976 she was a member of the Health Services Organisation Research Unit at Brunel University and a Tutor at the King's Fund College of Hospital Management.

David Frankel is a Consultant with HAY-MSL in London. He obtained his PhD at the London Business School for his work on advertising agency account group effectiveness.

Hugh Gunz is a member of the R and D Research Unit at the Manchester Business School, where he has been working on problems connected with the management of professional scientists and engineers in industrial R and D settings.

Anthea Hey has been a member of the Social Services Organisation Research Unit at Brunel University since 1969. She is a qualified social worker with extensive experience of field residential training and managerial work in Children's Departments.

Anthony Hopwood is Fellow of the Oxford Centre for Management Studies and Professor of Management at the European Institute for Advanced Studies in Management, Brussels. He is the author of *An Accounting System and Human Behaviour* and *Accounting and Human Behaviour*, and editor-in-chief of the international journal *Accounting, Organizations and Society*.

Kenneth Knight is Director of the Management Programme at Brunel University and is engaged in teaching and research on organisation and organisational change.

Peter McCowen joined Scicon as a behavioural sciences consultant in 1969 where his work led to the introduction of a matrix structure. Following further work as a consultant with the Tavistock Institute he now runs his own business.

Alan Pearson is Senior Lecturer in Decision Analysis and Director of the R and D Research Unit in the Manchester Business School. He is editor of the journal *R and D Management*.

Derek Sheane is an internal consultant working for ICI. Based at headquarters, he is engaged in examining and advising on issues of management and organisation, productivity and management response to changing economic and social pressures.

Introduction: the compromise organisation

The matrix phenomenon

Matrix organisation seems to have crept up on us while we were not looking. The name started to appear in the 1960s to describe a type of structure which, it seemed, already existed in various places, though up till then it had been called by other names such as project management or programme management, or had not been given a name at all. No one claimed to have invented it; no one tried to sell it as a great new management technique. Managements who adopted matrix structures perhaps did so rather uneasily, as a lesser evil, and kept quiet about it. Most of the management theorists at first ignored it or else mentioned it in passing as a logical third alternative to functional and product-based organisation and something of a curiosity. Most managers were unenthusiastic about it (most managers still are). In spite of this the use of matrix management began to spread in the late 1960s and early 1970s. It seemed to be a way out of a fix and an answer to some real needs.

One of these was a very basic, down-to-earth requirement faced by companies in the American aerospace industry. Donald Kingdon (1973), formerly of TRW Systems, tells us that, in order to be considered for government contracts, development organisations had to have a 'project management system' and it was actually necessary to submit organisation charts showing how the project organisation was related to top management. Most of these firms, however, already had a functional structure of specialist departments which they were reluctant to scrap. The alternative was to superimpose a set of 'horizontal' project groups over the 'vertical' functions. The resulting organisation in which individual engineers might find themselves working both in a specialist department under a department head and in an interdisciplinary project team under a project manager could be represented, as in Fig. I.1, by a grid or matrix – hence the name matrix organisation.

Kingdon saw matrix organisations as a compromise between two sets of needs: the customer's need for unified direction of the project to avoid having to negotiate with a series of separate functional managers, and the company's need for continuity as a viable, developing organisation building up its capability to handle future projects as well as current ones through the existence of strong specialist departments.

1

Functions or departments

		A	B	C	D
Projects,	1				
products or	2				
business areas	3				
	4				

Fig. 1.1   Matrix organisation

The desire to compromise between conflicting objectives and principles of management has been the keynote of matrix structures all along, the major compromise being between efficiency and flexibility. The pursuit of economies of scale through division of labour leads to concentration of resources, people and plant, into larger, more specialised units integrated by taller management hierarchies and fatter procedure manuals. Diversified markets, changing demands, economic uncertainties, technical breakthroughs and hybrid technologies involving the interlocking of dissimilar specialities all call for the quick and flexible response which only small integrated units can produce. The compromise is the project team, the product group or the business area team, with its membership drawn from the functionally specialised departments.

If the matrix organisation is a comparatively new development it is because the situations that have called it into being are themselves new: larger organisations with more complex interdependencies, tasks, techniques and knowledge changing faster than ever before and a changing social context. Such situations, as J. R. Galbraith (1973) has shown in a short, but most illuminating book, place enormous new demands on the systems for processing, distributing and using information within organisations, overloading the normal communication channels of the management hierarchy.

According to Galbraith there are four possible responses to such an information overload. The first he calls 'creation of slack resources', which simply means the acceptance that things will take longer and cost more, that inventories and budgets have to be increased, specifications relaxed and deadlines extended in order to cope with additional complexity. Secondly, the organisation may try to shed the load by decentralising into smaller, self-contained units. If this is possible without increasing costs then the apparent complexity must have been self-generated, arising from an attempt to agglomerate into large units activities that are not really inter-dependent and creating the 'diseconomies of scale' which some economists have recently been discovering.

A third possibility is to reinforce the hierarchy by investing in vertical information systems, such as on-line computing, which enable decisions to be taken more frequently from a more comprehensive data base. This can be very expensive and the experience with 'integrated management information systems', all the rage ten years ago, seems to have been patchy to say the least.

This leaves a fourth alternative, which is to multiply lateral relations that cut across the vertical lines of authority. Such relations create additional communication channels, relieving vertical overload and increasing the ability to respond directly at all levels to new data and changed circumstances. Galbraith lists a number of devices ranging from informal contacts between managers through task-forces and co-ordinators to a full-blown matrix organisation. Significantly, Galbraith sees matrix as the most difficult and expensive of these solutions: a last resort if all else fails.

Not everyone, however, has been as cautious as this. As awareness of the matrix option has become more widespread it has been taken up by some of the critics of present-day hierarchies, who see in matrix the organisational form of the future, an alternative to hierarchy more adaptable to rapid social change, more responsive to individual needs and closer to the democratic ideal.[1] It is easier to understand this perception if we link it to the earlier vogue of 'organic' organisation. In one of the most influential books about organisation ever to be published, Burns and Stalker (1961) argued that organisations operating in an uncertain and rapidly changing environment (in this case the electronics industry of the 1950s) were more likely to be successful if they were very loosely structured with flexible, undefined and overlapping roles, an emphasis on horizontal rather than vertical communications, on information and advice rather than instructions and on authority based on knowledge

3

rather than on rank. This type of organisation they called organic. Now in spite of the fact that it was obvious from their description that this kind of organisation had to be very small, in spite of the reported discomfort of the managers operating in it, in spite of the reported contentment of the members of the contrasting 'mechanistic' organisation and the latter's effectiveness in a stable, predictable setting, in spite of the fact that they described the organic and mechanistic structures as merely the extreme ends of a continuum – in spite of all this the organic organisation rapidly became an ideal advocated by the more radical, behavioural science-oriented authors, because it seemed to fit their model of 'self-actualising man' and the 'adaptive organisation' far better than the somewhat restrictive and formalised bureaucracies that most of us have to work in.

Unfortunately, however, experience has shown that it simply is not possible to run a large complex organisation by the principles of the small electronics firm described by Burns and Stalker in which the managing director refused to have an organisation chart and people had to 'find out' what their job was through continual interaction and negotiation with their colleagues. This is where matrix organisation comes in. On the face of it, matrix seems to provide a means of reproducing the more attractive features of organic organisation within the context of the large, differentiated hierarchy and this may account for some of the enthusiasm it has inspired.

The horizontal component of the matrix, the project team or product group, seems to be freed from the shackles of hierarchy and bureaucracy; it is seen as a peer group of colleagues able to short-circuit official channels to get together, pool their knowledge and get on with the job in hand. The junior scientist or engineer, sole representative of his discipline, enjoys the status and authority of special knowledge. Because team members come from different departments none can over-rule the other and decisions have to be made by discussion and agreement. Communications are open and intensive, relationships flexible and informal and response to external change can be rapid. Hence the matrix is seen as transforming the rigid, status-bound hierarchy, making it flexible, adaptable and democratic. Whether it always works in this way remains to be seen.

Tracking down the concept

But what, in fact, do we mean by matrix organisation? Definitions range from the very wide – so wide that it is hard to find an organisation that is not included – to the very specific. According to the widest definition any organisation involving lateral groupings or relationships in addition to a vertical authority structure qualifies to be called a matrix. This interpretation leads logically to the view that matrix organisations are ubiquitous and we have simply failed to recognise them. 'It is obvious to anyone who has worked in even the most stringent hierarchy, or bureaucracy, that any organisation is really a matrix or mixed model with multiple-channel communication.' (Kingdon, 1973, p.5)

The definition proposed by Corey and Star (1971, p.3) is almost as wide: 'A business organised by both resources and programmes which are integrated by means of co-ordination functions is said to have a matrix organisation.' This definition differs, however, from the completely universal one of Kingdon by making it clear that there needs to be some conscious adoption of a dual structure before we can talk of a matrix.

A more restricted definition is offered by Galbraith (1971) and is illustrated by a diagram which shows a continuum of relative influence going from a purely functional to a purely product or project structure (Fig. I.2). Matrix organisation is shown occupying a narrow central band, in which members are subjected to dual managerial authority and the relative influence of functional and product managers on decision-making is in balance.

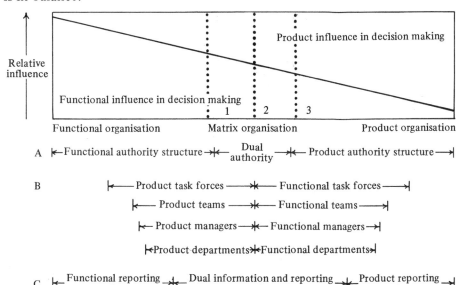

Fig. I.2   The Range of Alternatives (from J. R. Galbraith, 'Matrix Organisation Designs', *Business Horizons*, February 1971)

As our aim in this book is to explore a relatively uncharted field it would seem wise not to prejudge the boundaries of our undertaking too precisely by tying ourselves to a very specified definition at the outset. Definition works by exclusion; can we at this stage afford to exclude some of the more borderline types of management structure as 'not truly matrix'? If we do, we run the risk of missing some important insights. Yet there is an obvious need to fix a starting point and indicate the area under discussion, if not by drawing unequivocal boundaries, then at least by focusing on some of its central landmarks.

Most of the discussions of matrix seem to concentrate on the issue of dual, or multiple, authority. What characterises a matrix organisation for many people is the fact that it puts individuals in the position of having two bosses.[2] To my mind this begs some of the most important questions such as 'what is a boss?' Here the dangers of definition are immediately apparent. If we define matrix in terms of dual management we thereby prejudge one of the key issues, namely that of the respective role relationships along the rows and columns of the grid. In some reported cases, for instance where project managers are said not to have 'line' authority over the members of their team, the nature and origin of their power to influence the latter in order to achieve the project objectives becomes an important issue for the effectiveness of their role.

For this reason we might do better to focus in the first instance not on dual authority but on dual group membership. Let us say that one of the defining characteristics of a matrix organisation is that it contains members who are allocated to two groups. Most commonly, one of these is a specialist or functional department, the other a project team or product group. Other forms of grouping are possible of course, the next most common being geographical. What is important about these groupings is that both are work-related: their constitution is based on the need to get the organisation's work done. Membership of friendship or representative groups, in addition to a work group, is not relevant. They also have to be legitimate: informal groups or cliques are excluded. They tend to be ongoing; the existence of *ad hoc* working parties or occasional committees is unlikely to be seen as constituting a matrix though the difference is, admittedly, one of degree.

Still other cases need to be excluded. Heads of departments may also be members of a management team or board; thus membership of two groups, one of which controls the other, is not part of the matrix concept. Neither is the situation where one group contains the other. Often project groups are set up within a single department and where this is possible the complications of a matrix structure can be avoided, unless, of course, the project group includes members of several sections and these sections straddle more than one project. In this case we are back to an intra-departmental matrix structure.[3] Another situation which does not necessarily imply matrix working is that in which individuals divide their time between several projects: this is very common in consultancy groups. This becomes a matrix only if there is an attempt to organise by specialisation or discipline as well as by project as in the case described in Chapter 4.

One interesting point which emerges from these attempts to clarify the matrix concept is that the idea of dual leadership in some shape or form cannot easily be excluded from it. Almost all the group characteristics we have mentioned – work-related, ongoing, legitimate – seem to imply it, while, where one group controls or contains the other, unity of control is restored. But, as the exclusion of parallel projects shows, dual leadership seems not to be enough to make a matrix. There needs to be some difference in the organisational basis underlying the two groupings, e.g. common specialisation as against common objectives. Another clear indication that dual leadership is not in itself a sufficient basis for talking about matrix organisation is the very wide use of 'dotted-line' relations in organisations which few people would want to describe as a matrix.[4] These tend to be used where group membership is clear and single but where some other form of individual influence, jurisdiction or dependency has to be acknowledged. What distinguishes the matrix organisation is not merely the fact of two-way influence or authority but this fact in conjunction with a two-way division of labour, so that group membership as well as authority reflects conflicting principles of organisation.

By choosing in this book to focus on matrix *management* we have, in a sense, turned the precise definition of matrix *organisation* into a subsidiary issue. What we are concerned with is the whole range of approaches to multiple goal and multiple influence situations which organisations have to provide for. This has two implications. Firstly it enables us to refer to some organisation structures, such as 'project in function', which would not fall within any possible definition of matrix organisation, but which are clearly alternative solutions to the same problems and hence closely related. Secondly it emphasises the fact that we are concerned with process as well as with structure, with how matrix organisations can be made to work effectively as well as with their definition.

Is the matrix concept useful?

It is possible to argue that talking about matrix organisation is not very useful and might even do more harm than good. The argument might run roughly like this. The only precise and unequivocal way of describing an organisation structure is to describe the role relationships of which it is composed. This means defining each job-holder's responsibilities and his authority. What is he accountable for and to whom? What instructions may he give, to whom and about what? (Examples of role definitions of this kind will be found in Chapter 10.) But the matrix concept does not carry any clear meaning in terms of role relationships. Indeed, as we shall see in Chapter 10, a choice has to be made between different models of matrix organisation involving relationships being defined differently. Hence talking about matrix does not tell us anything precise about the organisation in question; at best it oversimplifies a complex situation and at worst it may confuse the real issues. For when all is said and done, we still have to decide exactly what the roles are going to be, so why bother talking about matrix in the first instance? Why not concentrate our analysis straight away on the role relationships to be set up?

The way in which the matrix notion can do positive harm, so the argument might go on, is by propagating the crude 'two-boss' idea. This is an open invitation to equate different types of authority and influence, disregarding the important differences between managerial roles, supervisory roles, co-ordinating roles and others. The result of setting up such ill-defined, confused and confusing structures is likely to be conflict, diffusion of accountability, individual stress and organisational pathology. I do not disagree with this second part of the argument. Rushing into dual-authority structures without pausing to define the respective responsibilities of each 'boss' may indeed be a recipe for disruption and instability.[5] The fault here, however, lies not with the matrix concept as such but with the failure to think through its implications in a concrete situation. Which brings us back to the first question. Why bother with the notion of matrix at all; why not just define the requisite role relationships which will meet the needs of the particular situation?

The view that clear definitions of role relationships are all we need seems to discount the importance for the manager of being able to think broadly, strategically, about the structure of his organisation. The decision, for example, to decentralise profit responsibility to product managers has to be considered at a more general level of analysis than the specific responsibilities to be written into the product manager's job description. Decentralisation may be a very broad and imprecise option, but it is an option nevertheless with known pros and cons and it sets the context within which individual roles and relationships can be sensibly defined.

The same applies to matrix organisation. Managers need concepts for thinking about organisation structure at a policy making level. At this level thinking becomes bogged down by specific role definitions or even generic ones. What is needed are broader concepts which can be related to organisation-wide objectives and problems.

Where an organisation is under pressure to develop new products or institute new services involving the collaboration of different functions and specialists, organisational options have to be considered. One of the broad options available is a matrix structure. Within this there are, of course, various sub-options which will result in differently defined individual roles. But unless the matrix concept is available a whole area of possibility may be overlooked. The idea of crossing lines of authority or influence may simply lack the legitimacy that would allow it to be considered seriously. Worse still, the logic of the situation may force people to engage in unofficial relationships which are not countenanced by the manifest organisation as it is expressed in charts and manuals. This puts them in a position where they can actually get into trouble for doing what the situation demands.

This points to another very useful function of the matrix concept. It provides a vocabulary for legitimising a whole range of organisational relationships which are often thought of as unofficial, improper and somehow undesirable and a framework within which different managerial strategies can be considered and compared. As soon as we are able to group different management structures used in a range of different organisations under a single generic heading, as is done in this book, it becomes possible to analyse and classify them and to provide a more systematic view of the structural options available. It also enables the operational problems managers are likely to encounter to be recognised and anticipated. A whole range of managerial experience is available for us to learn from, but without the label on the pigeon-hole marked 'Matrix' there is far less likelihood that this experience will be brought together and the common factors extracted from it.

Exploring matrix management

These uses of the matrix concept also define the aims of this book. The primary objective is that of exploration. If matrix does indeed represent a broad organisational option on the same level of generality as the functional organisation or the decentralised profit centre, it is, at the moment, a very cloudy one for most managers. For this reason our first aim is to find out what actually happens in practice. What types of organisation are using the matrix, under what circumstances, with what results? (The last is not always an easy question to answer.)

Part I is concerned with these questions. In Chapter 1 we survey available information on the application and spread of matrix management generally, while Chapters 2 to 7 present case studies and surveys of matrix management in six specific areas, namely research and development, manufacturing companies, management consultancy, advertising agencies, health services and local authority social services. The picture presented in Part I cannot be claimed to be comprehensive. We do not even have enough systematic information to be sure it is representative but, at the very least, it provides a sampling and a set of illustrations of current practice to enable us to base speculation and theorising on a reasonable spread of actual experience.

Our second aim is to supplement this picture of current applications of matrix management in the field with a theoretical account of the aims and objectives, or the organisational rationale, of this type of structure. If the matrix is to provide managers with a usable option, even in the most general of terms, they need to know which sorts of organisational problems can be tackled by matrix management and which organisational objectives it offers a hope of achieving. Under which circumstances would it make sense even to consider a matrix solution?

But we have already seen that matrix as a general choice is too indeterminate to be usable. To turn it into a specific option it is necessary also to distinguish between the different versions of matrix, the models from which a choice has to be made. What are their significant differences and what principles can be used for choosing between them?

Part II is concerned with constructing such a theoretical and descriptive framework. Chapter 8 raises the issue of the effectiveness of matrix management and tries to show that questions about the effectiveness of organisational forms require a special kind of second-order criteria, here described as 'criteria of structural effectiveness'. Chapter 9 uses these criteria as a framework for assembling a coherent theoretical basis for matrix management from the work of past and present writers and researchers on organisation in order to discover the circumstances and situations to which the matrix option may be appropriate. Chapter 10 tackles the specific question of the actual form of organisation to be employed by describing some major types of matrix structure and a number of variants related to them and by suggesting some criteria for choosing between them.

The operational problems of matrix management underlie the third aim of the book which is to present some current views and experience on how matrix structures can be made to work more effectively. What are the problems which have been experienced by people managing and working in these organisations and how can they be tackled?

Part III attempts to provide some answers. Chapter 11 reviews the main problems that have been reported and considers the reasons for them. Chapter 12 surveys some approaches to the effective management of matrix structures, starting with their impact on the internal distribution of power and going on to consider the importance of defining role relationships, of implementation strategies, of training and organisational development interventions and of management support systems. In Chapters 13 and 14 two of these approaches, organisation development and management information systems, are considered in greater depth. Chapter 15, finally, outlines the main questions about matrix management to which answers need still to be found and proposes a set of practical steps to managers planning to introduce a matrix structure. In conclusion it looks at the social trends which underlie the emergence of matrix organisations and considers the ability of people to work in them.

Summary

The use of matrix organisation has been spreading in the last ten years in response to a number of new needs. It has usually been a compromise between contradictory requirements and particularly between pressures for a responsive and self-contained project or product organisation and the greater efficiency, expertise and economies of scale of a functional or departmental structure. It has been seen as a way of dealing with greater environmental complexity and as a more democratic alternative to traditional hierarchical structures.

To attempt a firm definition of matrix at this stage might prejudge some of the key issues to be explored, but it appears that the concept must include membership of two or more work groups as well as some form of dual leadership, though the latter need not necessarily be defined in 'manager–subordinate' terms along both dimensions.

Against the objection that the matrix concept is too vague to be useful, can be dangerously misleading and should be replaced by a definition of specific role-relationships, we have argued that broad organisational concepts like that of matrix can be useful in thinking strategically about organisation structures, as well as helping to legitimise existing and necessary relationships. It can also provide a basis for comparing and distinguishing specific organisational models and for consolidating managerial experience of certain types of common problem. The chapter ends with a brief over-view of the rest of the book.

Notes

[1]   See Knight (1976), pp.117-18, for a summary of this literature.

[2]   For simplicity I will talk about 'two bosses', 'two groups' etc., when I mean 'two or more'.

[3]   In Chapter 10 the 'project within function' is described as a variant of one of the types of matrix organisation whose advantage is that it does away with the need for having a matrix structure at all. Its application, however, is limited to a particular type of situation.

[4]   But as indicated in Chapter 10, under the heading The Functional Overlay, there is now a tendency to transform dotted-line relationships into matrix structures.

[5]   The choice between clarity and deliberate ambiguity about the roles in a matrix is discussed in Chapter 12.

Part I

Applications

1 Areas of matrix management

Distribution of matrix structures: the sources

Exposure to matrix structures, and even willingness to believe in their existence, seems to vary considerably as between managers. Some find the concept inherently improbable and are convinced that when tracked down the so-called matrices will turn out to be perfectly normal management structures with a bit of functional authority or some line-staff relations thrown in somewhere. The other extreme is represented by the research manager who took me to task for implying that matrix organisation is an 'optional extra' – as far as he is concerned it is simply an inescapable fact of life. This variety of outlook owes much to the fact that matrix organisations are rather unevenly distributed; they are much more likely to be found in some contexts than others.

The aim of this chapter is to provide some kind of a map – though a rudimentary and incomplete one with many blank spaces – of the distribution of matrix management. The map does not show population densities, the data for assessing which are simply not available, but I suspect that they, are rather low, even in such well publicised areas of application as Research and Development.[1] The intention at this stage is to provide a background, to set the scene for the six contributions that follow, each of which looks at matrix management in a specific setting.

The main sources of information are in the literature on matrix organisations, much of which is represented in the bibliography.[2] This is supplemented from personal contacts and some information collected in a rather *ad hoc* and unsatisfactory way by circulating a short questionnaire to a management mailing list for course publicity. Although, for convenience, the latter will be referred to as a 'survey', the term is not really appropriate, the sample being neither random nor representative and yielding a very low response rate made up predominantly of respondents from organisations using some form of matrix, and hence probably an untypical minority. Nevertheless, it does provide us with statements from about 200 individuals in about 180 British organisations of various kinds who say they are using matrix management. Currently more detailed information is being obtained from some of the original respondents in a second stage of the survey, and while this information is far from complete at the time of going to press, some of the preliminary results emerging from it will be mentioned where relevant.

Antecedents

The ancestry of present day matrix management can be found in three managerial forms of long standing: project management, product management and functional authority. The most influential of these has probably been project management. In the literature the overlap between project management and matrix organisation is considerable, e.g. Janger (1963), Cleland and King (1968, 1975). If, as we have seen, the American government forced some of its contractors into matrix management by its insistence on a project organisation, the logic of this demand has been implicit in the management of interdisciplinary projects all along. Hence, it is perhaps not surprising that the chief incidence of matrix organisation seems to be in those areas where people are engaged in project work, particularly in research and development, consultancy, advertising and construction.[3]

Product management, in spite of its 'production' sounding title, has developed largely in the sphere of marketing, as is more evident from the alternative title of brand management, cf. Fulmer (1965), Corey and Star (1971). But in recent years there has been a tendency for product managers mainly concerned with sales promotion to give way to programme or business area managers concerned with businesses or market groups, and having co-ordinating authority over large areas of manufacturing companies. Thus some of the large scale company-wide matrix organisations have probably evolved from product management.

Functional authority, staff authority and dotted line relationships have, as noted in the Introduction, been a common feature of the management scene which would hardly rate as a form of matrix management. One of the surprises of our survey was that quite a few people are now prepared to describe the organisation of specialist services, such as accounts, personnel and training, in matrix terms, and in at least some of the cases the use is justified by a conscious attempt to establish dual group membership as well as some form of two-way accountability for the specialist. But this use of the matrix concept seems, as yet, to be absent from the literature.

16

Research and development

A large part of the literature on matrix concerns project management in research and development organisations. Much of this relates to the American aerospace industry and particularly to contractors working for NASA and the Apollo programme.[4] But there are also descriptions of matrix organisations in various other research and/or development settings both in Britain and the USA, including semi-conductors, telecommunications, guidance systems, flight simulators, construction research and toiletries.[5]

Our survey confirmed this picture, with R and D organisations forming the largest single group of matrix users, including twenty-one independent research/development establishments or companies and twenty-five in-company R and D functions. Matrix organisations in R and D receive more detailed consideration in Chapter 2.

Manufacturing industry

A few large international or multinational companies seem to have introduced company-wide matrix organisations, overlaying product or market groups over functional and/or area groupings. These include most of the divisions of ICI (see Chapter 3), aircraft manufacturers, like Lockheed and BAC, and American multinationals like ITT and Dow-Corning.[6] Research into the organisation of multinational companies has shown a tendency for such firms to develop either through product divisions or through a regional structure towards matrix-forms involving both types of structure overlaid on each other, in what can become very complex multi-layered organisations.[7] Perham (1970) lists a number of large American firms but makes it clear that most of them only use matrix management for some parts of their operations.

Our survey included twenty-three manufacturing companies in the UK reporting matrix structures which extend over the whole company. Company-wide use of matrix management was concentrated particularly in oil and chemicals, aerospace, electronics and computers. Altogether almost 100 manufacturing companies made use of matrix management in some part of their operations, major areas of use being R and D, management services and management of special projects, but with a fair scatter of applications in other areas such as marketing, production, engineering, accounts and personnel.

Consultancy

Both management consultants and other types of consultancy organisation seem to gravitate naturally towards matrix structures, because of the project nature of their work and the need to bring to bear different specialities within given projects. The literature refers to a number of examples,[8] but the fullest published account of such an organisation is probably that by Peter McCowen in Chapter 4. Twenty-four consultancy organisations returned our questionnaire: all of these use matrix management, and in about half the cases the matrix extends over the whole of the organisation. In addition, many internal consultancy units within large companies are run on matrix principles. A number of industrial training boards involved in consultancy and advisory work also use partial matrix structures.

Other industrial, commercial and service organisations

Our survey showed the use of matrix in a variety of non-manufacturing enterprises, for instance in the construction industry, in distribution, transport, communications, broadcasting, insurance and banking. Additionally, David Frankel makes the case, in Chapter 5, for regarding the typical account organisation of advertising agencies as a form of matrix structure.

Higher education

Our survey showed that a number of business schools, polytechnics, regional management centres, and similar establishments have adopted matrix structures. The usual form taken by these structures in higher education involves disciplines or subject areas on one axis and courses on the other. In some of the examples on which more detailed information has been obtained, the courses constitute the main administrative units, and provide the basis for recruitment and staffing levels. Subject units constitute resource groups for courses and may be concerned both with developing the knowledge of staff members by providing opportunities for contact, exchange of ideas, and support for further study, and with developing the subject resource itself by achieving a balance of specialisation, sponsoring research and developing teaching methods. Sometimes courses are grouped together into larger units, called departments or schools; the latter term is also used for groupings of related subject areas. A further version is a faculty structure, in which a faculty comprises a matrix of course running departments and their teaching resources grouped in subject areas and/or schools.

At present our information is not complete enough to assess the extent to which such structures have been adopted; in the universities their application seems to be extremely limited, with a very strong predominance of academic departments representing individual disciplines, but in polytechnics and other non-university colleges, where in the past courses rather than disciplines have often provided the basis of the administrative units, the tendency to place more emphasis on academic development seems to have encouraged the growth of matrix structures.

Local government

Local authorities, like other types of organisation, have been growing larger and more complex as they have had new demands and pressures placed on them. They have recently been subjected to some major reviews and re-appraisals, leading to large scale reorganisations. One of the first explicit statements in favour of considering matrix structures for local government came in the very influential 'Bains Report' on the management and structures of the new local authorities.[9] This, like Greenwood and Stewart (1972), based itself particularly on the programme planning and implementation structure then being developed in Coventry,[10] and on ideas which had already begun to be widely discussed in local government circles.

The Bains report envisaged, as one option to enable the development of a more corporate or global approach to the management of local authority services, the constitution of committees of elected members concerned with programmes rather than individual services, and supported by inter-disciplinary programme area teams of full-time officers. This approach is largely project-centred, and descriptions of the Coventry structure show that the primary tasks of such teams are the planning of programmes of work on the one hand, and the implementation of projects on the other.

The impetus towards this particular type of matrix management, which seems now to be developing in a number of authorities, is the increasing perception that there is a need to relate the management of local government activities to outcomes for the community, either in terms of specific groups, such as the elderly or deprived children, or of types of provision, e.g. housing accommodation, leisure, rather than to the specialised inputs provided by traditionally highly autonomous professional departments. The programme teams are intended to provide a horizontal overlay on the parallel vertical hierarchies of the departmental structure.[11] In practice, however, one often finds that major, supposedly 'horizontal' programme areas bear the same names as, and are dominated by, the vertical service departments, e.g. Education, Housing, Social Services, though some of the projects within these areas may be interdepartmental ones. At the same time some genuinely inter-professional programme areas seem to be emerging, like transportation, recreation and public protection.

Local government also has two other characteristics pushing it towards forms of matrix management. One of these is its inescapable geographical basis, and the other the distinction between departments concerned with service delivery, such as education, and those concerned with the basic resources used, namely finance, personnel and land. (Because of the statutory basis of so many local government activities, administrative and legal expertise probably counts as a fourth 'resource'.)

The geographical basis is reflected both in the electoral system, with councillors being elected by localised constituencies, but serving on programme or service committees, and in the fact that service provision usually requires some form of sub-division into areas or districts, with local offices being set up in different parts of an authority. The pull between functional specialisation and area provision has led to departmental structures which often have matrix characteristics. An explicit three-dimensional matrix in an education department which overlays programmes, functions and areas is documented by Mann (1973); Hey describes, in Chapter 7, how forms of matrix management are developing in social services departments, and similar developments are probably taking place in other departments. In addition, some local authorities seem to be developing more embracing area structures, related both to programmes and to services. There are also some indications of attempts to overlay resource management structures on the basic professional services. At present this is probably most common in relation to committee administration and legal services, and to accounting functions. Other developments may well be under way. Local government organisation in the UK is at present in a state of some ferment and we may have to wait a few years for the new structures to become clear and explicit.

Other public sector organisations

Some forms of matrix management found in the British National Health Service are discussed in Chapter 6. It is probable that other forms of matrix exist in health services in the UK and elsewhere. One instance, known to the author some years ago, was that of a psychiatric hospital which applied the therapeutic community concept. Because of the strong emphasis on work in the community outside the hospital, the organisation was based on the dual principle of professional disciplines (medical, nursing, social work), and geographical areas, with each of the major areas being served by an inter-disciplinary team.

From some recent conversations it appears that forms of matrix structure, with or without the name, may well have been in use in military organisations for a considerable time. A characteristic example of this, it is suggested, would be the deployment of support and service units within a formation, e.g. engineers, artillery, etc. within a division. While a careful distinction is normally made between operational and technical control, it is suggested that the authority of the 'technical' commander can under certain circumstances include the rapid redeployment of a support unit, thus creating a dual authority situation.

Scattered applications in other public sector organisations, in addition to nationalised industries, research establishments and management consultancy groups, were shown by our survey. They include certain government agencies, a number of industrial training boards and one voluntary agency, and tend to be related either to management of projects or to the co-ordination of similar functional specialists across organisational boundaries.

Summary

Matrix structures have originated from three rather diverse sources – project management, product management and functional authority – and their distribution over different types of organisation is fairly uneven. The chapter presents an overview of the occurrence of matrix organisations in research and development establishments, manufacturing companies, management consultancy groups and other industrial and commercial organisations. In the public services, matrix management is found in some areas of higher education, health services, some government agencies and perhaps in the military sphere, and matrix structures are also starting to emerge in the rapidly changing sphere of local government organisation.

Notes

[1]   Gunz and Pearson (Chapter 3) found 25 per cent of their British sample of R and D establishments had some form of matrix; Marquis (1969) in his American sample found a proportion of about 30 per cent.

[2]   This statement relates to the British and American literature. I have only recently become aware of a substantial German literature on the subject, unfortunately too late in the day for it to be included in this book. A few useful German references appear in the bibliography – Bernhard (1974), Brings (1976), Schemkes (1974), Thom (1973). There may well be important references in other languages of which I am not aware.

[3]   The use of matrix management in construction projects is little mentioned in the literature – there is a brief reference in Perham (1970). However, our survey showed at least five large UK construction organisations using matrix structures.

[4]   See e.g. Rush (1969); Sayles and Chandler (1971); Steiner and Ryan (1968); Wilemon and Gemmil (1971).

[5]   Lorsch and Lawrence (1972, pp.210-57); Bergen (1975); Videlo (1976); Paskins (1977); Hobbs (1969); Wilkinson (1974).

[6]   Corey and Star (1971); Brooks (1970); Goggin (1974).

[7]   See Brooke and Remmers (1970), Stopford and Wells (1972), Davis (1973).

[8]   E.g. Lorsch and Lawrence (1972, pp.258-65); Ludwig (1970).

[9]   *The New Local Authorities: Management and Structures*, HMSO, 1972, p.61.

(The) combination of the traditional 'vertical' structure and the 'horizontal' inter-disciplinary working group is known as a matrix form of organisation. This type of structure can operate both through programme teams giving advice and service to committees concerned with the general administration of particular programmes, or at the more detailed level of execution through teams working on specific projects. The membership of teams at either level can be amended or supplemented, new teams can be set up and existing ones disbanded as circumstances require. Herein lies one of the great advantages of this matrix system of management. By its nature it is flexible and adaptive, unlike the rigid bureaucracy which we suggest that it should replace.

[10] A summary description of this structure appears in Greenwood and Stewart (1974, Reading 16, pp.235-48). Stewart also gave evidence to the Bains Committee.

[11] Other examples of this type of development are the 'Programme Boards' set up by the Greater London Council (Peterson, 1972) and the project management system set up for New York City in 1968 (Lorsch and Lawrence, 1972, pp.291-309).

2 Matrix organisation in research and development

H. P. GUNZ and A. W. PEARSON

Introduction

Many attempts have been made to get some common agreement on the definition of research and development, or more particularly on its component parts. People have, for example, tried to distinguish between research which is basic or applied, short or long term, and defensive or offensive. In some cases such distinctions have proved to be useful, but in others they have confused rather than assisted those responsible for the management of resources in the area of science and technology.

In the same way attempts to examine the technological capability of an organisation by looking only at the R and D laboratory have often proved to be of little value because of the importance of technical work which is undertaken in other functional areas, for instance in design and production. In fact examples can be found of technologically oriented organisations in which there is no separately identifiable R and D department, and where the management of new technology is accomplished through the mechanism of project teams set up for the purpose at a particular point in time and disbanded on completion of their task. Co-ordination in such cases may well take place through a small secretariat (cf. Milner, 1972). In addition a considerable amount of the R and D department's time in many organisations is taken up with support to other functions, e.g. production and marketing. This work may be very short term, it is obviously applied, but it does not follow that it cannot be of a very fundamental nature; it may even be exploratory rather than a specific type of activity.

We would therefore argue that the characteristics of R and D will differ between organisations and that they will also vary between projects, even within a single organisation. If we accept this we must expect to find that the types of structure found in practice will differ widely, and that there will not be a simple form of matrix organisation suitable for all purposes. For this reason we propose to examine the requirements of different types of situation we have identified from a study of about forty R and D organisations, to see how different types of structure assist people to manage their activities. The organisations operated in a wide range of industries in both public and private sectors, including chemicals, foods, textiles, engineering, metallurgy, construction and packaging. They varied greatly in size from a formulation laboratory for a small speciality chemical manufacturer with a technical staff of eight, to one of several corporate laboratories of a large multinational group, with a staff of over 1,500. Of these 25 per cent had matrix structures of some kind. Before looking at their structures, however, we need to explain what we think makes R and D different from other parts of the organisation.

Characteristics of research and development

The key to understanding what R and D is, and why it should be of interest to students of matrix organisations, lies in the contribution it makes to the organisation of which it is a part. R and D groups may vary greatly in character from firm to firm, but in some way they are concerned with new activities, products, processes and directions, and translating intangible concepts, ideas and theories into specifications, tangible products or processes. As Woodward (1965) points out, this does not necessarily mean that R and D is at the start of the workflow. Particularly with small batch or unit production, its task may be to translate orders obtained by marketing into specifications which production can work up. The result is that R and D organisations, and the people working in them, tend to differ in a number of respects from other parts of the firm, and this has important consequences for the kind of organisational structures adopted. Some of the most important differences are:

(a)   the nature of the R and D organisation's inputs and outputs, which are connected with ideas and knowledge rather than with physical materials;
(b)   the level of analysis of the organisation's work: much of the kind of knowledge built up in R and D of the firm's products and processes is irrelevant to the needs of other functions of the firm;
(c)   the innovative content of the work, which tends to be much more concerned with relatively finite work on new things than that going on in other parts of the firm;
(d)   the R and D organisation's greater need for professionally trained staff who concentrate on the firm's technology rather than on its day-to-day operating exigencies;

(e)   the time scales of the work, which tend to be long; partly for this reason and partly because of the innovative nature of the work uncertainties tend to be high as well.

These distinctions will vary in their individual significance from organisation to organisation, and they begin to point to ways in which R and D organisations might be structured. As a number of writers have pointed out, the matrix can be very applicable to R and D organisations or high-technology industries.[1] Before turning to look at the kinds of matrix we have found in practice, we need to examine more closely some important needs which any R and D organisational structure must help meet. This will yield clues to why different organisations adopt different variants of the matrix, why sometimes they work and sometimes they do not.

Requirements which R and D organisational structures have to satisfy

We have argued so far that R and D differs in a number of significant ways from other parts of the organisation. These differences generate a number of requirements which the R and D organisation's structure has to meet. They vary in their relative importance from firm to firm and in some cases conflict, so that the process of organising is generally one of achieving workable compromises. We shall return to this point in a moment.

There are at least nine important requirements posed by the nature of the task and the needs of the organisation of which R and D is part. The organisational structure must:

(a)   mobilise resources to meet new work goals, which may not fit neatly into the existing structure, and which may involve many different parts of the organisation;

(b)   ensure that all of the knowledge and technology available in the organisation is fully used so that, for example, the 'not invented here' syndrome does not spread and that compartmentalised, over-specialised thinking does not miss fruitful approaches to solving problems;

(c)   allow efficient communication between those who need to be involved in the work in hand, which may mean that each new project needs a new network of communication pathways to be set up;

(d)   maintain some sort of boundary to protect staff from too much buffeting from elsewhere in the organisation; the longer time scales of R and D work can mean that other parts of the firm may get impatient at apparent inaction and try to influence the work,

perhaps to its benefit, but perhaps, on the contrary, to its detriment;
(e) allocate efficiently and harmoniously the laboratory's resources, particularly common facilities, between everyone needing them,
(f) ensure specialised skills and knowledge are maintained and developed;
(g) cope with the particularly difficult problem of handling the flow of ideas that constitute the inputs, throughputs and outputs of the system, a great deal of which information is in the heads of the staff and not in a form that is easily codified or otherwise handed on to someone further down in the work chain; Burns' comment 'the mechanism of technological transfer is one of agents, not agencies', has a lot to do with this particular requirement and the following one (Burns, 1969);
(h) allow for a smooth and efficient transfer of the output of R and D to whoever is to implement it;
(i) be compatible with the structure, rules, procedures and so forth of the larger organisation of which R and D is part.

It is clear, therefore, that an R and D organisation has many conflicting demands placed on it. The problems this may cause in practice can surface in all kinds of ways, not the least of which is constant reorganisation. As we shall try to show, by illustrating with examples known to us, the matrix structure is capable of handling many of these conflicting demands, and, because of the way in which it allows lateral groupings to form and disband, of institutionalising reorganisation. Our experience is, however, that the pure matrix, as represented in Fig. 2.3, is encountered only rarely, and in the next sections we shall look at the variants that are found in practice and the reasons for their adoption.

Matrix organisation in R and D: two models of leadership

Matrix organisations – the term covers a very wide range of forms of structure in R and D – seem to fall into two classes, each reflecting a set of assumptions about how people ought to be managed. Within the two classes variants occur as responses to the varying influences of the needs that we have described above. We have called the two main classes the leadership matrix and the co-ordination matrix. The assumptions that seem to lie behind these two styles are in many ways peculiar to R and D, reflecting the views R and D workers have of themselves in two contrasting ways.

The leadership matrix

(a) assumes that people tend, if left to themselves, to pursue their own ends, which are likely to be related to their professional, specialist goals, and need galvanising into working on the organisation's goals;

(b) needs drive, motivation, leadership towards tackling the task goals;

(c) needs the leaders of the lateral groups to have influence and status, so that they are not dominated by the heads of the vertical groups;

(d) requires the lateral groups to be cohesive and strongly motivated to solve the problems necessary to complete the task.

The co-ordination matrix

(a) assumes that people are rational and objective, work to achieve the organisation's goals given adequate information, and balance priorities in a way acceptable to all concerned given the facts of the situation;

(b) needs everyone to be kept informed about what is happening to the task, and when they will be required to do what;

(c) entails the leaders of the lateral groups being seen as co-ordinators, with the most complete knowledge about what is happening to the task and what it will involve in due course, and able to influence events by signalling to the vertical authorities when the project deviates from its plan;

(d) has lateral groups which only need to be a set of nominated individuals concerned with the completion of the task, each knowing who else is involved. The nominated individuals are primarily members of the functional groups, and project work is secondary to their functional work.

Like all models, these caricature reality, and elements of each type of matrix can be found within single organisations. To quite a large extent organisational structures are subjective entities anyway, depending for their existence on how individuals in the organisations interpret their individual positions and their relations with their colleagues (McKelvey, 1975). Our experience is, however, that one reason for the difficulty which R and D organisations have with matrix organisations is that the underlying differences between the co-ordination and leadership matrices often have not been recognised, and that, for instance, people think they are operating in one while the dominant ideology of the organisation reflects the assumptions of the other. In one laboratory, for instance, project leaders are encouraged to see themselves as operating in a leadership matrix, and the more ambitious take to this naturally. The organisation is, however, very hierarchical and, unless the project leaders are old, experienced individuals (which many are not, since often scientists are put in charge of a project shortly after joining the firm) they stand no chance of influencing the heads of the vertical groups. The rules and operating procedures in fact are those of the co-ordination matrix, and the frustration that results tends to come from this confusion. These points, with some added comments, are summarised in Table 2.1.

Table 2.1
Two models of matrix organization

	Leadership matrix	Co-ordination matrix

The model assumes people	– tend to pursue their own goals:
	  – professional
	  – specialist
	– need galvanising into working to project goals
		– are rational, objective
		– act predictably on adequate information

Role of project leader is to:	– motivate team to work to project goals	– keep everyone informed about:
		  – project status
		  – when their contributions will be needed

Consequences for project leader:	– needs status, authority – gets action by personal authority, influence, negotiating skills	– is co-ordinator, with most complete information on status and future needs of project – gets action by signalling deviations from plan
Consequences for functional manager	– scope of authority, responsibility limited by project needs	– must consider project needs in conjunction with needs of functional activities
Consequences for project team	– must be cohesive group – functional activities interfere with project work	– meetings of nominated individuals – project work interferes with functional activities

In our survey of the ten organisations which operated some kind of matrix, roughly half of these appeared to have co-ordination and half leadership structures. In the next section we look at examples of these, to see how they work out in practice.

28

The co-ordination matrix

In its simplest forms the co-ordination matrix is as shown in Fig. 2.1. Individuals in each of the departments, or groups of them, are nominated to serve on a project as the latter progresses through the system, and each project is overseen by a project co-ordinator who may be found in a separate group devoted to this activity, or who may be a member of one of the departments through which the project progresses. The people with a current interest in the project meet regularly under the chairmanship of the co-ordinator, but do not form a day-to-day work group. The workflow, of course, is rarely as straightforward as shown. Activities will go on simultaneously in some departments, and as more is learned about the problem it may be referred back to earlier departments in the sequence. The co-ordinator's role is to remain aware of what is happening in each department, alert departments to what will be expected of them in due course, and act as a focal point for dealings with customers and other outside bodies. Examples of this structure can be found in the pharmaceuticals industry, where the range of outside bodies that have to be involved in the process of developing a new drug is very high. In the construction industry, a similar structure has proved particularly successful, having survived in very much its original form since 1966 (Hobbs, 1969).

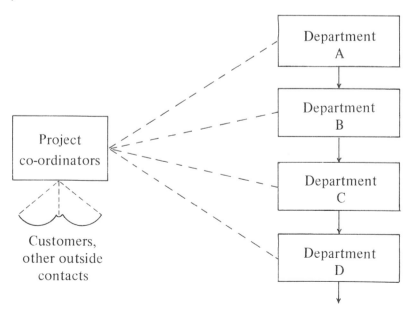

Fig. 2.1   Simple co-ordination matrix

As well as acting as a focal point for communications, the co-ordinator also acts as a kind of repository for much of the information held in people's heads which otherwise would be lost if the work was constantly passing from one person to another. Typically he is a very experienced individual who has spent a long time working in the industry and knows a great deal about most aspects of the development process. In the example quoted in the previous section it was clear that the greater the company experience of the project leaders, the less trouble they had co-ordinating their projects. A key success factor is that the co-ordinators are often senior R and D professionals, usually with functional management experience, so that their word carries considerable professional weight with the R and D staff.

The project co-ordinator meets the organisation's boundary maintenance need by acting as the principal interface with the organisation's environment. Part of his information-processing job is to act as a filter so that the right messages get through from the environment, and others, which might interfere with the group's task, do not. This is obviously a highly ticklish job given the sensibilities that are so easily aroused when people feel that information is being withheld from them, but no less essential for that. In the example quoted from the construction industry, the greater part of the incoming work is routed through the co-ordinators. It is worth noting that the structure mirrors the way in which the industry works as a whole. Firms interact with the ultimate client via the architect, who is an independent operator, and in turn co-ordinate the work of independent sub-contractors. In this way, the concept of working for an individual to whom one does not formally report comes naturally – to some extent at least – to the kind of people working in the R and D centre.

Conflict-resolution over scarce resources remains a problem, however. Each project co-ordinator is responsible for one or a few projects, but when a constraint is run into, for instance in department C (Fig. 2.1), the co-ordinators needing that department to work on their problems find themselves in the position of suppliant to department C's manager. For most of the time they are able only to put the facts of the situation to the head of department C and hope that he sees it their way, so that day-to-day control of the process of balancing the claim of each project (which may or may not be an optimising process) tends to rest in the hands of the head of department C. He may or may not be guided by priorities assigned by some kind of vertical supervisory body, which might consist of all the vertical departmental heads or of their line manager. In the ultimate, the project co-ordinator can only signal his problem to the supervisory body, and it becomes the latter's business to decide how the scarce resources in department C are to be allocated. Even in those organisations which have been operating a matrix structure for a long time, we find that this kind of conflict occurs, but is handled to a great extent by the fact that the managers and senior staff know each other well (at senior levels staff turnover is low) and can usually sort out problems on the basis of mutual professional and personal respect. This is partly dependent on the size of the R and D centre: in a very large laboratory this kind of arrangement would almost certainly be unworkable.

Sometimes it is not so much the common facilities that are the pressure point as the amount of co-ordination needed between groups working on the project. In a pharmaceuticals firm, for instance, the situation is much as is shown in Fig. 2.2. In the early stage of drug development the degree of co-operation that is needed between a number of the departments is so high that just a project co-ordinator calling meetings every so often would seriously limit progress of the project. Project teams are formed in the early stages, therefore, allowing much fuller interchange of ideas and information between chemists and biologists than would otherwise be the case. In due course a product candidate appears, and vast areas of uncertainty and unclassifiable information held in people's heads can be sharply reduced. The project passes from the team who are conversant with all of this 'messy' knowledge into the simpler co-ordination matrix, under the control of a highly experienced project co-ordinator.

The structures described so far have taken a very cool, rational view of R and D staff. When R and D managers describe their organisations they usually start by talking about the need for efficient communications and co-ordination, but at some stage the problems of motivating people enter the picture. Sometimes it is the common resource pressure point that causes trouble, but more generally managers describe the problems faced by a project co-ordinator who has to try to get people, over whom he has no formal authority, as motivated as he is to get the project task completed. In principle, the more project activities can be programmed, other things being equal, the less each step in the project will be accompanied by a great deal of unstructured communication, negotiation, persuasion, and so forth. The 'ideal' project, after all (from the point of view of the project leader at least), is one in which every activity is predicted and scheduled and goes exactly according to plan, for then, after the initial plan is made, the leader hardly needs to do anything. As the project gets more complex, however, even the best-laid plans will tend to go increasingly astray, needing much more intervention from the leader. Also, the more innovative the project, or the closer the project to its

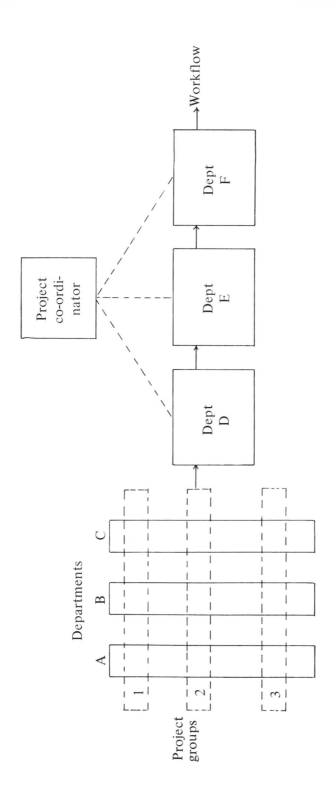

Fig. 2.2　Front-end matrix

innovative stages, the less rigorously programmable will activities be, once again demanding more intervention from the leader.

Our experience is that the co-ordination matrix is found in laboratories, or those parts of laboratories, where the proportion of programmed work is fairly high. As soon as the amount of information that has to pass from one part of the project to another in order to hold it together starts to take on sizeable proportions, it seems that managers become less happy to rely on the cool, rational assumptions about professional staff represented by the co-ordination matrix. More and more the structure tends to reflect the assumptions of the leadership matrix. The corollary of this is that a leader can minimise the day-to-day motivational part of his or her job by using planning methods as much as the job allows. It also seems likely, and we found it to be the case in practice, that as a project progresses the emphasis may well change from the need for a co-ordination matrix to need for a leadership matrix, or *vice versa*. We shall now look at some of the leadership matrices that firms adopt when they find the co-ordination matrix unsuited to their needs.

Leadership matrices

Fig. 2.3 shows a highly idealised version of what is usually understood by the term matrix organisation in R and D. The functional groups represent different disciplines or kinds of activities; the project leaders may themselves be members of functional groups or be independent. In one published example (Videlo, 1976), project leaders and division heads are separately identifiable and report independently to the same authority. The team membership may vary with the progress of the project, but typically a stable core will be involved from early until late in its development.

The advantages of this arrangement have been widely commented on. As noted in other chapters, such structures in R and D are among the first written about as matrix organisations, particularly in the United States. To recapitulate briefly, they are held to have the dual advantages of arranging for a body of dedicated individuals to be strongly committed to the project, while ensuring that they retain functional home bases which maintain their specialist standards and give them a place in the organisation to which they can return at the project's end. The boundary maintenance problem is solved by the project leader as described in the previous section. Because a recognisable group of people stay associated with the project throughout its life, essential knowledge and ideas are not filtered out, and for the same reason communication pathways unique to the project are automatically established. Most of the needs of R and D organisation, in other words, are nicely met.

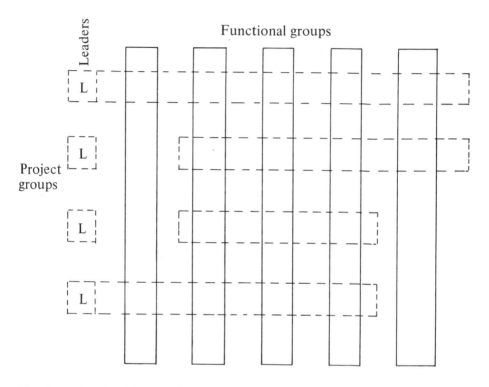

Fig. 2.3   Leadership matrix

Despite these powerful arguments, matrix organisations as depicted in Fig. 2.3 are not found very often in R and D laboratories, at least in the UK. They are sometimes found as variants, and we shall look at three of these shortly, but even so they have not generally been in use for very long, and have occasionally been abandoned because of the difficulties experienced with them. We do not mean to suggest that our experience is that the matrix is not a viable form of organisation for R and D, but rather that it does not solve problems in quite the way a cursory glance indicates that it might. There seem to be four problem areas, two of which are closely related. These are:

(a)   the influence of the project leader;
(b)   the relationship between the project team member and his line manager;

34

(c) uncertainties experienced by team members as a result of the fluid nature of the project teams;

(d) over-identification of project team members with the team.

The first of these, the influence of the leader, is inherent in the leadership matrix concept. In the co-ordination matrix we looked on the leader as someone who provides information to functional heads and others, on the basis of which they could reach the important decisions. Here, however, the leader is someone who himself is making important decisions. The team is seen to be much more responsible for running the project, and the role of the heads of functional groups becomes less that of controller and more that of ensuring that their staff are provided with the facilities they need. This immediately raises problems of the relationship between functional heads and the project team, embodied perhaps in the leader. The former can very easily see the latter representing a threat to his position, potentially usurping part of his authority, a problem which can be particularly severe when the matrix is first introduced to the organisation. An example of what may happen as a result was found in the laboratories of an oil company, where the functional heads rapidly ensured that they were also the project leaders, even though this was not the intention behind the new structure.

It is for this reason that many firms operating the leadership matrix find it important to have project leaders of similar status in the organisation to that of the functional heads, so that the day-to-day conflicts resulting can be sorted out between peers rather than on a suppliant/superior basis. It is also the reason other R and D managers often give when maintaining firmly that the matrix is unworkable. The director of a large laboratory serving a nationalised industry tells of the downfall of such a structure because of the inability of the horizontal and vertical managers to work together, despite, or perhaps because of, their equal organisational status.

These difficulties can be turned on their head in some instances by looking on the project leader's job as the first step in a managerial career. Starting from the proposition that leading a project is a very difficult task, for all of the reasons we have gone into, it is argued that such a job is therefore a very good training ground for future managers. It is not an unduly high-risk situation, for the individual is responsible for only the project, which has a finite lifetime and finite consequence, and if he turns out not to be a success he can easily revert to the position in his line department whence he came. Anyone surviving the experience may well be launched on a successful managerial career. There is some evidence that an important factor determining the success of a project is how big a step-up it represents for the project leader by comparison with his previous project: the bigger the step-up, the greater the chance of success (Rubin and Seelig, 1967).

The second problem area is perhaps less serious. A functional manager who numbers among his subordinates project team members has to try to cope with the situation in which he is apparently responsible for staff who themselves spend a great deal of their time working with, or for, others outside the vertical group. The difficulties can surface in a number of ways. Who, for instance, is responsible for appraising the performance of the staff, and administering pay and promotion as a reward for the performance, or lack of it? How much should the project member keep his vertical boss informed of what he is doing, and to what extent should the boss demand to be kept informed, and control the work of the subordinate? and so on.

So far we have been looking at R and D staff as rational people, and as political people. The third and fourth problem areas in the leadership matrix are concerned with social man, and are closely inter-related. One, in fact, is the opposite side of the coin to the other. Two examples may make this clear.

The research department of a chemical company, which differentiates more than many between research and development work, was formed some years ago into a matrix in which project groups formed lateral links between functional departments. The culture of the organisation was such that the functional managers had built up strong departmental loyalties among their staff, so that the latter did not find it easy to make the transition to working in multifunctional groups. The issue was sharpened from their point of view by the fact that both functional and project managers shared responsibility for individuals' performance appraisal. Project lives were of the order of up to two to three years, a time scale found in a wide range of industries. Despite this, staff felt that in that time their project leaders could not get an adequate idea of their abilities, so that their performance reports might well not be fair. Additionally, given the existing strong functional loyalties, they found it hard to cope with being members of relatively short-lived groups as well; the constant formation and dissolution of project groups was too unsettling for them. Taken together, these and other factors added up to a culture in which lateral linkages proved too difficult to sustain formally, and the matrix was abandoned. As we shall show shortly, there are laboratories which do not experience these problems despite having markedly shorter project lives. Our point is that each organisation has its own unique features

36

making broad generalisation about the kind of structures that are suitable for R and D hazardous.

A metallurgical laboratory operating a 'split matrix' (see Fig. 2.5), on the other hand, is worried that members of the project teams will get too involved and committed to the work of the team so that they will keep working on the project long after they should have cut their losses. Given a dedicated team of people strongly committed to the project, the management of the laboratory foresees situations where they may tend to let the project continue rather than tackle the unpleasant task of stopping the work and breaking up the group.

To put these last two problem areas in perspective, however, it is worth quoting the case of an organisation in which project lifetimes are short and staff very committed to their projects, but the system is felt to work very well. An R and D consulting firm, in which projects are typically much shorter than in the chemical laboratory cited above, operates a system in which members of its staff are responsible themselves for bringing work to the firm. When they do, and show that the work is worth doing, they form a team from however many vertical groups in the organisation are necessary, selecting the individuals they want. The firm finds this works very well for a number of reasons. Projects get completed quickly because there is an individual and a team closely involved in and responsible for the work. A kind of natural selection determines staff appraisal, for those individuals who are much in demand do well, and those who are not are equally obvious both to top management and to themselves. The staff who thrive in the system stay, while those who do not leave.

Leadership matrices, then, are powerful devices for meeting many of the needs of R and D organisations, but they raise great problems as well. Individual organisations show a great tendency to select from the matrix concept those parts that are useful to them, with the result that there are many variants in use, tailored to the needs of the particular firm. We shall now look at three of these, and give reasons for their adoption.

The simplest variant could be called the localised matrix. One of the laboratories of a multinational company is divided into a number of areas of activity, each interacting little with the others. Within individual divisions the work tends naturally to break down into 'disciplines' (not necessarily corresponding to the traditional academic disciplines) which, in the original structure, tended to work relatively independently despite the management's efforts to improve co-operation. In order to correct this, matrices have been formed in a number of divisions, encompassing the disciplinary groups of that division, with project groups overlaid laterally. Initially only one sub-division was organised as a matrix, on a trial basis, and gradually other divisions are adopting the idea as it can be seen to be applicable to their individual needs.

The matrices can be even more localised. For instance, a research institute has formed matrix organisations for individual projects. The principle is illustrated in Fig. 2.4. The institute was commissioned to develop a novel item of equipment, and a first look at the problem suggested four possible approaches which could be tried. There were, as well, something like four specialisms that were relevant to the problem. A matrix was formed consisting of the four relevant specialisms divided laterally into four project groups, each project group taking one of the possible approaches to the problem. Four individuals were selected, each an experienced practitioner of one of the specialisms, to head both a vertical and a horizontal group. In this way the level of awareness of what was happening in each group was very high, and technical standards were uniform to each horizontal group, for each specialism.

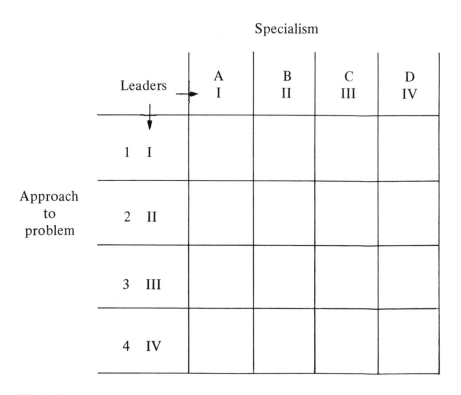

Fig. 2.4. Localised matrix

38

We have called the second variant the split matrix, illustrated in Fig. 2.5. The example is taken from a metallurgical laboratory in which the need was recognised to maintain certain specialist skills, which otherwise might be lost if the individuals concerned became deeply involved in the broader activities of project work. There was also the need to provide central analytical and other support services to the project groups. The organisation adopted is in two halves. The first consists of three pools of technologists from whom the project groups are drawn. The functional home bases are fairly vestigial; they remain responsible for salary and other routine personnel administrative matters but do not have a major responsibility for maintaining technical expertise. Much of the latter is vested in the second half, the support groups, in which are found all of the normal services as well as the particular specialists who provide consultancy services to the various projects. Members of the support groups do not normally become full-time project team members, although individuals may become closely involved in particular projects from time to time.

The third variant is illustrated in Fig. 2.6. We have called it somewhat inelegantly the back-end matrix, since it aims to deal with one of the most intractable problems of R and D management, the implementation of the laboratory's output. The innovation literature is widely agreed on the need for ideas and technology to be transferred by people, by means of face-to-face contact, and if possible with particular, influential individuals closely and continuously identified with the innovation's success or failure. In some way the R and D organisation's boundary has to be made semi-permeable, so that it is capable of both insulating and de-insulating R and D staff when appropriate.

Many approaches have been used to tackle this problem, and the back-end matrix, of which we have found examples in many industries, is one such. In the latter stages of the project a group is formed, encompassing such departments as parts of R and D, production, marketing and cost accounting, generating commitment among these groups and an adequate flow of information, and ensuring that the necessary extra unclassified knowledge is available to those implementing the project. Leadership of back-end matrices depends very much on the nature of the firm and the patterns of influence within it. If the firm is production-orientated it may well choose the production manager to lead the team, if marketing-orientated the marketing manager, and so on.

It is at this point that the distinction between matrix management in R and D and in the firm as a whole becomes somewhat vague, since we are entering the province of the product or business manager/co-ordinator who stands apart from individual company functions. This is entirely as it

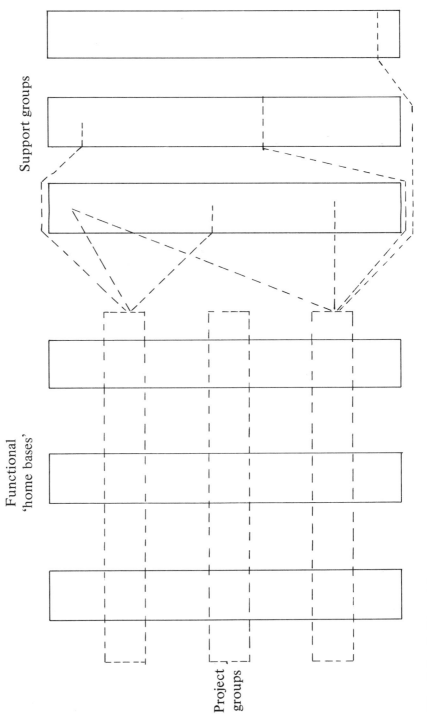

Support groups

Functional 'home bases'

Project groups

Fig. 2.5   Split matrix

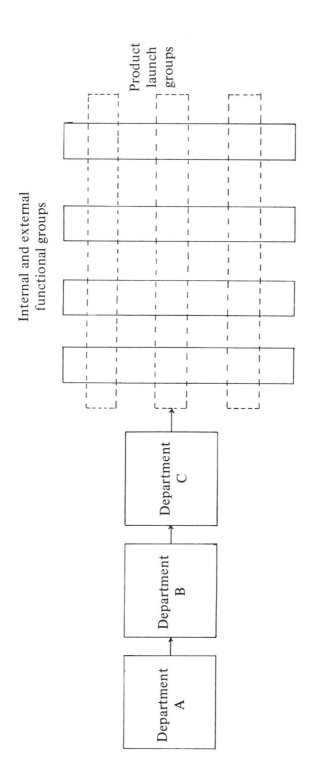

Fig. 2.6 'Back-end' matrix

should be, because the aim of the back-end matrix is to link R and D into the rest of the firm. Relationships between the parts of a business are usually a complex function of the nature of the market, the technology, the history of the firm, and so on. Given the variety that results, it is to be expected that the ways in which different firms choose to formalise the links between functions will vary greatly.

Conclusions

Many authors have recommended the matrix as a suitable, if not the ideal, organisational structure for R and D. Indeed the roots of matrix organisation, some maintain, are buried in the R and D world. In this brief survey of matrix organisation as it is found in the parts of UK industry with which we are familiar we have suggested that there seem to be two kinds of matrix, each reflecting contrasting views of how R and D staff need to be managed.

In earlier sections we have shown how, in individual cases, organisations have adopted matrix structures to meet particular needs. We have selected examples which, in our view, deal with problems shared by many laboratories. Explaining the logic of how a particular structure meets given needs is one thing, but explaining why the laboratory should go for a matrix structure at all is quite another. For every laboratory which has adopted a matrix there are many in similar situations which have not, even though they may have considered the possibility carefully. By the same token there are laboratories in similar situations which have adopted different kinds of matrix, or which have had quite different experiences with structures which are very much alike. There are many important general questions like these which at the moment we are unable to answer properly for two reasons.

Firstly, we lack the data. We think that the answers will emerge from detailed examination of not only the firm's technology and markets, but also its history, the way it rewards its staff and the kind of careers they have. For example, a common problem faced by laboratories in the 1970s is that many experienced great growth in the 1960s and cutbacks in recruitment and staffing more recently. This sequence of events has, among other things, spawned a generation of young researchers unhappily aware that their promotion chances are slender. Matrix organisation tends to be welcomed by such people because it gives them the opportunity of managing projects, but resisted by older middle managers because of the way their positions of influence are apparently undermined. Obviously, then, one cannot compare two laboratories contemplating matrix structures without allowing for the effect of age distributions and the career aspirations of the staff.

Secondly, we lack the measures of success or failure that are attributable to the structures. Some matrices have been abandoned and some appear to be operating happily, but since one never has an identical organisation with which to conduct a controlled experiment one can only guess, with the best evidence to hand, what would have happened if some other structure had been adopted. It seems to us that the best criterion for guessing at the extent of the benefits gained by an organisation from its structure is whether, after some years, the latter is still in use. Unfortunately, only one of the matrices we have described is more than five or six years old, and most are considerably newer, so that even here we cannot comment easily.

It has been argued in many places that the needs of R and D organisations are best met with the aid of a matrix structure. Our feeling is that such generalisations are dangerous, particularly given the wide range of activities that come under the R and D umbrella. In our experience different organisations adapt various aspects of the matrix concept to their own purposes, giving rise to many variations of the matrix structure in practice. There is no evidence, in our view, that any one type is most likely to be suitable for all situations.

Summary

Many writers have traced the root of matrix organisation back to the Research and Development world, and in this chapter we examine some of the reasons why the matrix is still found to be useful there. R and D groups vary greatly in character from firm to firm, but in some way they are connected with new activities, products, processes and directions, and translating intangible concepts and theories into specifications, tangible products or processes. R and D, therefore, tends to differ from other parts of the firm in a number of ways, including among others the nature of the inputs and outputs, the level of analysis of the work, its innovative content, its time scales and uncertainty, and the kind of staff it employs. A laboratory's organisational structure has, as a result, to meet a wide range of conflicting demands which the matrix concept is often well suited to handle.

Despite this, in a survey of forty R and D organisations in the UK we found that only a minority actually appears to use matrix structures.

Those that do seem to fall into one of two broad groups, which we have called co-ordination and leadership matrices, and which differ in the assumptions they make about the way in which technical staff should be managed. The co-ordination matrix appears to be well suited for situations in which a variety of relatively programmable contributions must be co-ordinated and the urgency of the work is not over-riding. When these conditions do not all apply, the leadership matrix appears to be useful as a means of galvanising project team members into working towards task goals.

Some of the problems appear to be associated with the influence of the project leaders, the relationship between the project team leader and his line manager, uncertainties experienced by team members as a result of the fluid nature of the project teams and over-identification of project team members with the team.

Although the matrix is usually drawn as a simple two-by-two arrangement, in practice organisations often adapt this to their own needs giving rise to a wide range of variants. Some of these have been described and it is clear that particular laboratories have selected those parts of the matrix concept that are helpful to their own particular needs. This often appears to have been an adaptive process and not one which has been consciously imposed upon the organisation, which is probably as one would expect in the R and D field, given the nature of the work and the characteristics of the people involved.

Acknowledgement

The work described in this chapter has been supported in part by means of a fellowship granted to one of us (HPG) by the Social Science Research Council, to which we are most grateful.

Note

[1]   For instance Lawrence and Lorsch (1967), Galbraith (1973), Miller and Rice (1967) and Kingdon (1973).

3 The company-wide matrix

DEREK SHEANE

Rather than concentrating on the detail of matrix design I will try to raise and discuss matrix organisation as an expression of fundamental forces in modern organisations. As with any 'new' development it is important to look at its historical role and consider basic questions like: Is this an early expression of how large complex organisations may be managed in the future, or is it more simply a sympton of decline and papering over an internal institutional crisis? Is it an expression of a new cohesion and a step forward in human social skills or is it a move towards fragmentation? I believe these questions are worth asking for the following reasons, at least in the context of ICI and some other UK organisations I have knowledge of. Firstly, there is significant managerial dissatisfaction with matrix as a way of working and a majority preference for simpler management systems. Secondly, it is hard to assemble any clear evidence that the use of matrix, *per se*, leads to an increase in performance or in human satisfaction. I can certainly see the use of matrix correlated with better performance, but there is usually at least one other factor to which people working in the situation attribute improvement, for example, 'we work the "two boss" system here but things really took off when the board got their business policy crystal clear and won our commitment'.

Thirdly, I have a strong feeling that what is called matrix organisation has always existed in human and social history and wonder to what extent the widespread ignorance and disinterest of modern behavioural science in such disciplines as social philosophy and history could account for the belief that matrix is new or novel. The evidence for suggesting this as a serious point is that many ICI managers can point to matrix systems being used several decades ago and my own findings that these forms of organisation are used unconsciously and naturally in some small firms where multiple roles, two and more bosses, and flexible use of resources across functional boundaries occurs spontaneously.

To deal with these questions and explore underlying forces, let us look at the use of matrix in different ways based on ICI experience. As a starting point let us examine the history of matrix management in ICI.

The history

Nobody knows when matrix structures first developed, but it was clearly long before the traditionally accepted birth of the matrix in the American aerospace industry. Some clearly rudimentary forms existed in 1947, as for example in the General Chemicals engineering department and elsewhere, but whether they were effective was another matter – certainly they were not understood or given a name. As with all organisations the matrix form evolved before it was described, largely because we appear to need a researcher external to the system to put in the necessary effort to develop a description.

This process of evolution had several aspects which occurred at different times. Some of these were:

(a)   matrix has existed as an informal structure since at least 1947;
(b)   problems of matrix have had an impact on effectiveness since at least 1947;
(c)   the concept of matrix has been recognised overtly since about 1965;
(d)   the problems and dynamics of matrix have been understood only since 1970;
(e)   since at least 1975 there has been significant concern about the usefulness and benefits in some situations.

The setting up of a matrix structure has occurred when three dynamics conspire to produce a multiple catalytic effect. These are:

A trigger incident
A dilemma requiring management
Some form of facilitation

The trigger incident is usually related to the external environment in which the organisation operates. It often takes the form of a traumatic impact on the business which generates a clear pressure for a change in structure or behaviour. This environmental pressure creates energy for doing something.

Examples of triggers are the need to cope with a vastly increased capital construction programme, a market development failure in introducing a product into some overseas territories, or the need to deal with declining business or other significant commercial upsets. Often anything that produces the need to use resources more effectively has this trigger effect, particularly if that part of the organisation cannot fall back on traditional strategies like more resources, self-contained units or investment in highly improved information processing systems.

The dilemmas requiring management usually revolve around two or more needs of the business which are interdependent but related inversely, so that the achievement of one restricts the achievement of the other: a situation of internally conflicting aims. Typical examples of dilemmas include the production of a specialised product for an individual market versus the need to maximise the use of the productive capacity, or an emphasis on detailed end-user market knowledge versus emphasis on a highly sophisticated product technology.

The form of facilitation can be based on the attributes of a person. Examples are: an ambitious man producing strong autocratic leadership; a change in leadership from one person to another who looks at the business with different priorities and beliefs; a change in leadership style; and occasionally a mature individual taking a personal risk which requires him to tolerate a great deal of uncertainty about his future in the organisation. Another significant form of facilitation is the situation where individuals in the problem situation just decide to set up informal systems to cope with pressure: this is a spontaneous emergence of a structure that perhaps later becomes institutionalised.

The social and business context

In the early 1960s there was a challenge to the basic belief that the company's success depended on its technical success. This led to greater emphasis being put on the company's ability to market. It occurred because the company had to take a more international view of markets; cartels and monopolies were becoming illegal or breaking down and the practice of dumping products on a marginal cost basis in order to utilise capital fully was extending in many aspects of the chemical business. This led the company to seek ways of improving its market effectiveness by increasing its orientation to specific markets or by an in-depth concentration on a narrow band of products. This required either a business area or a product group organisation to be laid over the traditional functional organisations with the result that there was a noticeable increase in the use of matrix organisations.

The second central intervention was all the work on social change programmes such as WSA and SDP which arose from the same commercial trigger as the previous one.[1] This focused on the problems of improving productivity. There was a deliberate aim to produce a cultural change in which greater participation in problem resolution should occur and to facilitate this there was a trend towards getting problems into the open more readily and facing up to an effective management of the issues concerned. This cultural change was not limited specifically to the productivity issue but led to other organisational problems being more readily raised, and in particular the basic problems of operating a matrix became recognised overtly.

If this is the history, what is the underlying dilemma? There are of course several but let us look at a key dimension.

Big is strong but small is beautiful

It is hardly surprising that groups and individuals want to walk on both sides of the street at once: in this case to attempt to combine the traditional benefits of size (technological scale effect, experience curve effect, dominant market share, etc.) with the virtues of developing small cross-functional business teams that identify with the total process of invention, manufacture and commercial exploitation. Looking around the organisation you can see many examples where a preference for self-containment or fairly complete decentralisation allowing a manager to have his 'own business' has been thwarted by the existence of an expensive functional resource that must, *ergo*, be shared with other businesses. This can occur within a product division or, with even more complexity, between two divisions. In this type of situation I feel sure the resulting matrix is a compromising or optimising system that tries to combine the benefits of size and identification with a smaller unit. Seen in this light it is a half-way house between centralisation and decentralisation. Matrix used in this way can be seen in at least three levels of ICI:

    (a)   Intra-functional
    (b)   Divisional
    (c)   Corporate

Intra-functional

An example is engineering, where the use of project teams for building a new chemical plant serves to optimise the three dimensions of time, cost and technological performance. By constructing a cross-functional, interdisciplinary team a mechanism is developed whereby different inputs of skill, knowledge and experience are co-ordinated in order to achieve the project goals. The inputs are the various forms of engineering: chemical, electrical, civil etc., the host organisation (the production unit), outside contractors, finance and personnel. When this works well the potentially conflicting goals of 'within budget' and 'on time' and 'make sure it works' are achieved.

Ideally if the mechanism of cross-functional collaboration is a flexible one then alterations in emphasis or changes in the hierarchy of goals, of performance, time and cost can be accommodated. It is here that the skills of role and behaviour flexibility are much in demand and an ethos of living with ambiguity and 'making life up as we go along' is required. Of course a key problem in a situation such as this is the definition of authority and responsibility. This is handled differently depending on circumstances, but with projects where expenditure is significant and high performance essential two processes take place. Firstly, individuals from the different disciplines and functions involved may go through an extensive off site 'getting to know each other' programme where attitudes to work together on the project are discussed: for example, leadership style, how work is to be planned, how priorities will be set or changed. Secondly, the project group may complete some form of responsibility charting where the type of authority and responsibility any individual or department has in relation to any key decision or issue, is discussed, agreed and written up.[2] This becomes the group's set of role and behaviour 'contracts'. This can be displayed for all to refer to and is ideally in a form where modifications can easily be made as experience develops over the life span of the project. In this way identification with project and the traditional benefits of functional management are combined.

Another intra-functional example is the way maintenance engineering resources can be allocated and deployed in a works. Here the problem can be expressed as sharing out scarce resources in the face of conflicting demands. For example, the service manager will have the usual situation of, say, three production units all wanting resources yesterday and have no clear method for deciding priorities. Again this is a situation where varying degrees of matrix can be used, the fullest expression being a team with members from all units that meets regularly to agree schedules and priorities. In relation to our theme the 'small versus big' issue is symbolised by the conflict experienced by the manager of a production unit where he can be torn between what is best for his unit (small) and what is best for the whole works (big). For the engineering manager who is providing the service the question is: what do I optimise? My resources? Production unit A or the whole productive capacity of the works? In this example you have some of the typical problems found in a matrix:

1   Uncertainty about who is the boss: for instance in the case of the team member he may wonder who he should respond to, the engineering service manager who is his functional boss or the leader of the cross functional team that decides priorities for the total works?

2   The emotional problems of identity: for the individual member it is a question of thinking through his group identity. Is my loyalty to be given to the works priorities or is it still the engineering department? Or can I be loyal to both: be a full member of both systems and somehow manage the conflicts in role and loyalty when they arise?

Divisional

The outstanding example at this level is what is termed a business area matrix. Within ICI there is no fixed pattern to typify how a business area matrix works and it is more useful to picture the different types on a continuum that shows the degree to which a business is self-contained and has all its own resources under its control.

1 Self-contained	2 Mostly self-contained	3 Mixed	4 Full matrix
The business has all resources under its control, e.g.:	The business shares only one or two important resources e.g. a pooled sales force	The business shares a significant number of resources with other businesses:	The business has only an integrating role or department. All key resources are shared between different business interests
production research sales finance personnel, etc.		production research sales	

One can see how as one moves from 1 through 2 and 3 towards 4 life becomes more complex and co-ordinating problems multiply. Position 1 is a simple situation with no matrix characteristics, whereas 4 is the opposite. Positions 2 and 3 are intermediate examples: obviously there can be more steps or gradation than are shown in the diagram. The point of the scheme is to show that in terms of organisation design management has a choice. As one would expect, a large multi-national like ICI will contain almost every possible example on even quite a refined continuum. Increasingly, different parts of the organisation are seeking out what is most appropriate. For example, one large division which manufactures a wide range of

50

products and serves a wide range of customers and markets has been a full matrix for some time: i.e. the division has four or five major businesses that share common resources. Accordingly it is an internally complex division that exhibits many of the typical matrix problems. Indeed the matrix is a source of real concern to those operating it and this, in part, is responsible for one of that division's businesses being increasingly run on a self-contained basis, i.e. a shift towards position 2.

On the other hand another business within the division is run very much on a full matrix basis with two formal lines of authority (business and function), multiple group membership and a whole array of complex integrating devices: an integrating department, a permanent cross-functional team and personnel reward systems that examine manager performance from both a business and functional point of view. Also the presentation of business information reflects the reality of matrix and its key conflicts: for example the presentation of sales and cost data to sharpen up the effects of selling different products in different territories at different prices. Other divisions would have analogous situations where they manoeuvre often by trial and error, to find the best mix or degree of matrix for their operation.

In practice this choice is not always exercised, for, as inhabitants of any large complex organisation know full well, conformity is prized more than difference and within any division's management system there is always the pressure to organise different businesses, and therefore tasks, in much the same way (management levels, grades, information systems and so on). The choice revolves around characteristic trade-offs most of which are expressions of the benefits of size (strength in bargaining, use of assets) versus the attractions of self-containment like market flexibility and employee identification and commitment to a business.

Here it is important to point out that it is not just a straight fight between big and small. There is another dimension: that of simple versus complex. In terms of management 'big' is not so much of a problem on its own: it is when complexity (issues, different vested interests, markets etc.) is added to size that potential unmanageability is a real consideration. Indeed a small complex organisation like a hospital will prove to be a greater managerial headache than a large, simple, vertically integrated manufacturing firm. So simplicity as much as, if not more than, smallness is the sought after characteristic that makes self-containment a desirable state of affairs.

When managements select positions further along the continuum and in particular as they approach the position of full matrix, part of the designing process is a conscious desire to balance efficient use of resources with a flexible response to the outside world. The difficulty in achieving this is well documented and has never been more apparent than now, when pressures on productivity collide head on with market orientation. These incompatible (but not unmanageable) pressures put more and more tension into matrix organisations and tend to disturb the power balance between interests, much prized by advocates of matrix, by pushing the management system to optimise on one rather than two dimensions. For example, it might seek self-containment and thereby optimise identification with the business and flexibility of response. A different type of response to the pressure would be to return to a more functional emphasis, thereby optimising the specialist or functional contributions. As the power balance swings back and forth due to environmental change, this is reflected internally. Indeed it is often suggested that the ability to change the balance between functional and business influence on affairs without major reorganisation is a key benefit to be derived from matrix systems.

In summary, then, ICI divisions use a variety of matrix types and the particular variation used will tend to reflect the business and social circumstances, though it is not always easy to produce a good fit especially in those situations where a high value is placed on internal conformity.

Corporate

So far we have only talked about attempting to reconcile two dimensions of management: business or project with function or specialism. But as we proceed to a larger scale of analysis, the total firm, matters become more complex and three dimensions of organisation require attention: product, function and geography. As a firm moves from being international (exporting from UK) to being multinational (making and exporting in different countries) then the conflict between product profit and territorial profit appears. Behind this simple statement lies a hierarchy of issues, some clear, some subtle that require extremely high governmental and organisational skills: at one level, the allocation of capital between different geographical or national territories, at another the influence of UK versus overseas companies on a whole range of issues affecting product policy. Add to this the reality of the different economic performances of different nation states together with varying social and political climates and you can see how the modern corporate executive needs something approaching superhuman design and operational skills in the field of matrix organisation.

Thus at corporate level in some cases a matrix is used to integrate the conflicting demands of product, function and territory. This makes the multidimensional organisation a reality with executives having to design and operate very complex cross boundary mechanisms. An example would be an international business area consisting of UK division representatives (product), head office planning and finance (function) and overseas company heads (territory) reporting to a corporate policy group. Ideally, this mechanism is designed to include key interests, and appropriately balance power relations in order to create constructive conflict that will optimise total business performance. This corporate example is matrix writ large and is on the leading edge of sophisticated management systems. Indeed the honest assessment is that ICI is unclear about the extent to which these mechanisms can always work and accordingly frequent review of their effectiveness is a way of life, and of course a route to improvement. The issue of 'small is beautiful' versus 'big is strong' with which I started is still very real here, in the sense that the company is trying to create islands of business and social identity amidst a sea of complexity.

Some fundamental problems

Looking at the operation of the matrix at different levels in ICI it is now possible to summarise common problems. There are three key ones that most managers would recognise as real.

Firstly, because the purpose of a matrix organisation is to cope with two or more opposing needs which co-exist, it necessarily has built-in instability. At one end of the spectrum there is pressure towards formation of a series of small units, each specialising in a particular part of the environment and being self-contained in terms of resources, whilst at the other there is a pressure towards a simple functional organisation. There is always a pull towards one or other of these polarities which can vary with time. It can develop for 'good' reasons such as an effective adaptation to a changing environment or for 'bad' reasons such as an internal power struggle which has been won or lost.

Secondly, there is a marked contrast between a simple hierarchical organisation and a matrix. In a simple hierarchy the siting of business decisions, the rules for the resolution of conflict and the development of basic business philosophies, and the power to reward and punish come together appropriately in one place or role. However, in a matrix most decisions are expected to be taken at the operational level at which business areas and functions interact; constant escalation of decisions to the 'cross-over point' can make the system unworkable.

Thirdly, the people who work within a matrix are subjected to a considerable increase in uncertainty and ambiguity because they have to hold multiple jobs and multiple responsibilities compared with a more traditional organisation. The central problem for the individual is therefore one of managing the personal dilemmas created by membership of different groups: the dilemma of responding to conflicting value priorities.

There are three other problems I would identify as an observer of the system as it operates but I could not guarantee industrial managers or indeed external professional colleagues would agree. These problems are cultural in that they reflect the past and current UK socio-political environment, not ICI management.

Firstly, there is a quest for simplicity. In some cases this verges on over simplification. This is symbolised by an interlocking set of demands for simpler management structures, smaller units, and less interference from external agencies whether government or others, and above all a rejection of having to manage more conflicting demands.

Secondly, there is a quest for leadership. This is symbolised by a demand for clear and unambiguous policy, by a covert request for a harder, tougher approach to affairs. It is also seen in a very marked and definite tendency towards centralisation of decision making in matters of capital investment and personnel policy.

Thirdly, there is a quest for identity. Some people in managerial roles are now quite unclear about their future, their identity and their basic reason for existence. As a reaction to this there is a tendency to be more cautious and to withdraw behind boundaries of function, specialism or role.

I believe all of these tendencies are in conflict with the climate and context necessary for the operation of matrix organisation. If this is true, then what can be done to counter these tendencies?

Implications for matrix management

In the first place the nature of the role of the chief executive and his immediate top team has to change. The balance of the use of their total time needs to change in a way that puts more emphasis on setting up and improving the effectiveness of the systems and ways of working rather than detailed involvement in business decisions. This means that the top man's job becomes less about making business decisions, and more about making decisions about how decisions should be made. There has to be a well developed sense of purpose and direction in the matrix as a whole to provide a framework for resolving business dilemmas and the approaches to resolving them.

At the same time much effort is needed to develop a greater clarity of people's roles, their complexity and an ability to be clear which role is being played at a particular moment. For example, it is possible for a person to have a role as a maximiser of the effective use of a plant or other assets (or as a maximiser of the effective business operation of a market area) coupled with a role of contributing to international policy formation of the product his plant produces (or business handles). This sort of situation produces role conflict which can lead to supporting a policy decision which is not optimal for the organisation in order to solve a problem in the person's plant role.

The basic organisational systems such as the information system, the reward system, career development system, etc., which have a major impact on the relationship of people to an organisation will require redesigning in order to reinforce the ways of working required and the dual responsibility situation implicit in a matrix. The reward system, considered in the widest sense, is very dominant in affecting attitudes and behaviour. A system is required by which the top team can demonstrate their understanding of the complex nature of people's roles, demonstrate what they value in each aspect of the role and the balance desired between the various aspects, and set standards of performance in each aspect.

Conflict resolution and dilemma management must be by open confrontation and problem solving backed by the use of authority rather than the smoothing out of conflict typical in a bureaucracy. There is no evidence inside or outside ICI of a matrix working effectively when conflicts are ignored and when a 'don't rock the boat' motto prevails.

A top management view

It is useful to examine the views of an individual who has actually had extensive experience of making the matrix work. I asked Robert Malpas, an ICI main board director and recently Chairman of ICI Europa, the question: What have you learned from the way ICI and other companies manage or fail to manage matrices?

His first point concerned objectives. Matrices work better if there is clear overall objective setting to which all other objectives can be linked. For example, what is the market ambition, the growth ambition? If this is explicitly done then all can participate. But so often it is only done by the principal power centres because they do not want to dilute power: sharing objective setting does this. Also they do not want to share information.

Malpas also draws attention to behaviour. Matrices do not work if behaviour is immature, for example, failing to face up to conflicts. Matrices will not work if there is a lack of role-definition in terms of divisions, departments and individuals. Indeed, the ability of individuals to play different roles is central and a key managerial skill is ability to define roles explicitly.

Finally he draws attention to management accounting. Present accounting methods and procedures for presenting information do not help. If profit is the motive then the presentation of information should harness this. For example, management information systems may reflect fiscal requirements and management structures only. What is needed is presentation by product and territorial market. This makes the conflict in the matrix explicit. It also turns destructive conflict into constructive conflict where the overall aim of the business is being optimised.

Looking to the future Malpas sees a natural process of development. Matrices will mature into federations – so it is matrices for the young but federations for the mature. As our large organisations develop they will increasingly manage the dilemma between 'big is strong' and 'small is beautiful' by developing federal structures. This will demand mature behaviour of the highest order, in particular managing power, conflict and multiple roles.

The way into the problem is therefore threefold: better strategic thought and objective setting, better behaviour as a result of clearer role definition, and better management accounting. All of these can be taught.

Conclusions

It would be too easy to set out to prove matrix is 'a good thing' so I have been deliberately critical of its use and sceptical about its future. But to round off the discussion I will try and summarise its operation in ICI.

The concept has been in use for several decades in ICI. The forms of matrix used vary a lot and successful application has been correlated with management's fitting the matrix design to the business and environment need. What is 'right' is what is appropriate. There is considerable variation in management's reaction to its use. Some find it an abomination, others a fundamentally improved method of organisation. While individuals may disagree on the usefulness of matrix, all agree that it is difficult to operate.

56

These difficulties are, I believe, related to deep underlying cultural values and personality factors such as a belief in simplicity and centralised control.

But in spite of the reservations I have expressed and other disadvantages managers might mention, many individuals in managerial and staff roles still believe we should carry on learning and developing the skills associated with matrix management, in particular the skills of multiple group membership and the lateral management of power relations and conflict.

I see no way forward if we cannot manage complexity, for a belief that man can manage his modern institutions as simple hierarchies is surely a vain hope. But to think we can achieve effective organisation-wide matrices or federations or other forms of complex organisation without questioning underlying historical and cultural values would be a still more vain hope.

Summary

Matrix organisations have developed in ICI over the last thirty years under the influence of various environmental changes, particularly an increased marketing emphasis, which have led to the need to balance conflicting pressures. One of the central themes has been the conflicting benefits of size and self-containment.

Examples are given of matrix structures at intra-functional level. At divisional level organisations range from completely self-contained 'businesses' to a fully two-dimensional business area matrix, trying to balance productivity against market orientation. At corporate level there is a development towards three-dimensional structures, subject to continuous review and modification. Problems identified include the potential instability of the structure, owing to a tendency to move towards one or other polarity, the difficulty in resolving conflicts hierarchically, and the personal dilemma of conflicting priorities. These problems lead to a constant tendency to look for greater simplicity, more forceful leadership and a clearer identity.

The implications for matrix management include a change in the role of the chief executive and his top team, the need to develop clearer roles and to redesign information and reward systems, and the need for a confronting style of conflict-resolution. A main-board director confirms these views, drawing attention to the importance of clear objective setting, role definition and presentation of accounting information.

Acknowledgement

The sections on the historical development of the matrix and the definition of fundamental problems make use of a working paper on matrix organisation by A. V. Johnston, Central Personnel Department, ICI.

Notes

[1] The Weekly Staff Agreement (WSA) and the Staff Development Programme (SDP) were company initiatives aimed at improving productivity and work relationships. The former is described in Roeber (1975).
[2] This process is described in Chapter 13.

4 Introduction and operation of a matrix organisation in management consultancy

PETER McCOWEN

Introduction

Scicon (Scientific Control Systems Ltd) is a management sciences and computer consultancy which has operated a matrix for the last seven years. The company, owned by British Petroleum, has during the last fifteen years conducted a wide range of assignments for British government, industry and commerce, as well as an increasingly wide range of work in Europe, the Middle and Far East. Work ranges from providing advisory services on management sciences applications and the use of computers to the installation of complex computer and communications systems. By 1969 the company was facing a number of problems concerned with adapting the supply and allocation of consultancy staff to rapidly changing products and markets while at the same time meeting the individual career needs and aspirations of the staff. It was becoming increasingly difficult to staff projects so that they matched the rapidly changing demands for computer services and changing patterns of work. A matrix structure was developed to meet these changing needs and to provide a basis for new methods of managing staff and the relationship with clients.

The original intention in 1969 was to add behavioural sciences to the company's range of consultancy services. However, it was recognised that the company had its own problems so it was agreed that a demonstration of what could be done for the company should be given before starting work for clients. The process of identifying, diagnosing, and interpreting the main staff and organisational needs took place over a period of nine months. Implementation of changes leading to the matrix took place within the next three months.

Background

The company, formerly known as CEIR Ltd, had grown rapidly during

59

the 1960s and had acquired a reputation as both problem solvers and providers of management sciences and computer applications to government and industry. The company had previously enjoyed a fairly informal management style but by the late 1960s was experiencing problems of size, having grown to 400 staff, and an increasingly complex range of activities. Senior management had adapted to these pressures by creating new divisions to cope with each major new computer development or area of application. The addition of each new division was causing problems of reporting and control and some reorganisation of the structure was becoming necessary approximately every six months. The result was that the consultancy side of the company was operating eleven different divisions, arranged in a hierarchy (Fig. 4.1). People complained about the longer reporting lines, the loss of personal contact with senior management, and insufficient information about the company and its affairs. What had previously been an informal and sharing management style was being experienced as increasingly formal and remote. The company was also losing staff at all levels at rates which threatened the development and progress of work. Senior staff were being attracted away to other computer organisations being set up in the late 1960s, often with American financial backing. Middle and junior staff were leaving because of uncertainty about their own immediate career prospects when in fact their existing skills and experience could readily have been used within another division.

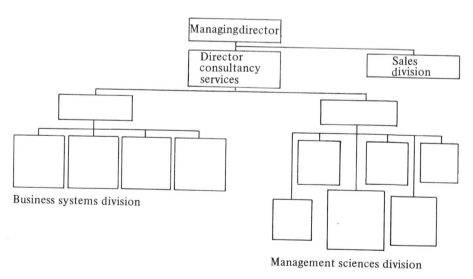

Fig. 4.1   Scicon's organisational structure before the introduction of the matrix

60

By 1969 these problems were of sufficient concern for senior management to invite me to mount a special study. This was the first time that the company had used a behavioural scientist not only to study its own problems but with the eventual intention of including behavioural sciences in the range of services available to clients. Preliminary discussions with senior management revealed some of the operational problems they were experiencing in co-ordinating the present structure and anticipating future developments. Some managers were also experiencing conflicting priorities in managing their present divisions. They had a wide range of responsibilities and were almost totally responsible for exploring, selling, staffing and managing projects within their own sector of the market. Conflicting priorities arose because of competition for their own time between ongoing work and developing and selling new projects. A project nearing completion required concentration and extra effort; however, this was also the time when, to maintain staff utilisation, most effort needed to be put into ensuring new work. This raised directly the problem of the appropriate size of a work group: teams of more than twenty people were proving difficult to manage by one person, given the range of skills of staff and the rate of change of projects. The other problem facing senior management was how to adapt the company to cope with the rapidly developing applications and markets for computers. It was proving increasingly difficult to anticipate not only new trends but also the rates of growth of new markets.

An attitude survey was conducted within the first months of the study to explore the issues that were concerning management and staff. The survey also served to create an exploratory framework within which the company could begin to examine some of its problems. The urgency of some of these was highlighted by the departure, within six weeks of the start of the study, of a group of senior staff to set up a rival organisation. The results showed that a high proportion of the consultancy staff were mainly interested in the technical aspects of their work; also that they got on well together, and that many were ambitious to share in the developments that were occurring in the computer industry. High levels of importance and satisfaction were expressed with many aspects of ongoing work. In contrast to this, high levels of importance but dissatisfaction were expressed on the various items concerning the allocation of work and guidance given about careers. Although of less importance, dissatisfaction was also expressed with the internal company procedures that concerned the control of work and knowledge of what was going on.

Questions about management style showed that staff felt their divisional managers gave them plenty of freedom in how they did their work. On the other hand, they also felt that they were given little guidance. An important finding was that many staff had no idea of the criteria by which their performance was being judged.

Staff ambitions and perceptions of future opportunities were shown by answers to questions about interests and prospects. Most importance was attached to having greater variety of work and having more responsibility for projects. Both of these possibilities were seen as likely within the next one to two years. Although some people were dissatisfied with the present variety in their work they were reasonably hopeful that their expectations for both variety and responsibility could be met. Private discussions had already revealed the disillusionment felt by some staff when their expectations had not been met. Some of these expectations had been based on insufficient information either about the company or about their own abilities.

Areas of moderate importance and also seen as attainable within the next one to two years included more responsibility for managing people, and opportunities to develop new computer applications and new techniques. Moderate importance was also attached to more responsibility for forming policy, and opportunities to work in Europe and further afield, but both these options were seen as unlikely within the company. The latter finding was indicative of the need for more communication in view of the active plans for European and overseas development being pursued by senior management at the time. Few people were interested in having more administrative responsibilities although this was seen as fairly likely, and few were interested in more responsibility for selling, marketing, teaching, lecturing, or in the possibility of working in Britain but away from London. The overall picture was of a strong interest in the technical aspects of the work with the expectation that more variety and responsibility could be realised, and a lack of interest at the time in the more commercial aspects of the company.

The approach to change

Each stage consisted of collecting information, diagnosing needs in discussion with the various interest groups, and exploring possible areas of solution. No change was introduced until the full implications had been worked out with, and accepted by, those who were likely to be affected. I certainly had no original conception that there would be changes in management roles or in organisational structure, nor did senior management. Each change was designed to meet the needs that had been uncovered so far without pre-empting any alternatives for the future. The attitude survey and the discussion of the results in a series of open meetings to which all staff were invited had raised a number of issues for the company in ways that were not considered too emotive and were not challenging any particular interest groups.

The first change was the introduction of training in the human skills of conducting staff appraisal interviews. This training was designed to meet the need expressed by staff in the survey for more guidance about careers and clearer guidance about the criteria being used to judge their performance. It was also designed to help managers and project leaders overcome difficulties some had expressed privately about confronting people whose performance was less than satisfactory. To do this we ran two-day training courses consisting mainly of role-playing based on case material incorporating typical problems being encountered in the work. The record keeping system was designed afterwards once the interpersonal skills and target setting aspects had been fully understood.

The other main pressure experienced by managers and senior consultants was the problem of organising so as to meet the ever-changing demand for the company's services and skills. This was becoming an increasingly complex problem and carried with it the risk that mistakes could be costly in terms of wasted effort or lost opportunities. Correct anticipation and organisation could put the company several years ahead of its competitors and enable it to become established in new and potentially lucrative markets. An internal market research study raised questions such as: How quickly would real time computing take over from batch modes of operation? What was the anticipated demand for 'turnkey' contracts, and how could these be financed? What was the likely demand for very large data handling systems by commerce and industry? Did particular techniques, such as numerical control, have a significant market potential? Other questions concentrated on evaluating areas of market potential. To what extent was government likely to commission outside contractors for its ever-widening use of computers? What were likely to be the software requirements of the computer manufacturing industry? What importance ought to be given to developing overseas business?

Initially the market research study had raised more problems than it solved since it revealed more alternatives and more areas of uncertainty than had been recognised. During the next couple of months a series of meetings took place with the twenty most senior managers to try to evaluate and structure the alternatives. There was much initial discussion of who would be responsible for particular territories rather than any recognition of the need to separate management roles. It was only towards the end of this period that it gradually began to be recognised that a separation of roles into resource managers and project managers might enable the company both to manage its existing work more effectively and to be better at anticipating future trends. The basis for a matrix structure had been created amid general relief that much of the uncertainty of the previous two months had been resolved. The new structure divided ongoing project work according to the main client groupings, namely industry, government, and computer manufacturers. Resource areas were divided up according to major areas of application and included business systems, turnkey and real time computing, management sciences, and consultancy (Fig. 4.2).

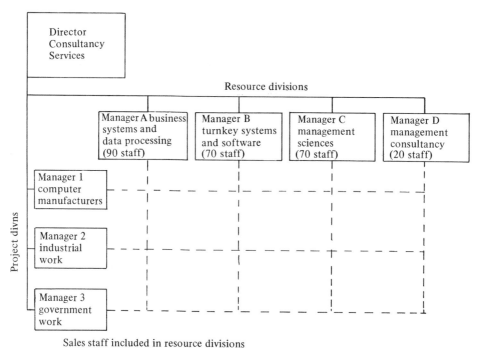

Sales staff included in resource divisions

Fig. 4.2   Scicon's matrix (as introduced in 1969)

There followed another month during which the roles and areas of work of the resource and project managers were clarified and managers assigned to the new jobs. Because most of those affected had been involved in the preceding debate there was widespread understanding of the reasons for separating the two roles. The responsibilities of the *project managers* were the most readily implemented because of the obvious similarities to those of the existing project leaders. They included:

(a) agreeing the estimates of work effort and costs of all new projects with resource managers;

(b) planning manning levels and team structure;

(c) planning, implementing and monitoring the progress of projects with regard to work completed, quality, costs;

(d) on-the-job training;

(e) responsibility for obtaining extension or repeat business from existing clients.

Resource managers were each allocated one of the identified areas of computer development and market application. Their responsibilities included: (a) maintaining staff utilisation; (b) recruitment; (c) overall training and development of staff; (d) developing and bringing in new work (other than from existing clients); (e) R and D activities relating to computer applications in their field.

As the two management roles emerged it was agreed that they should have distinct but balanced authorities. It was not intended that either type of manager should have control over the priorities of work of the other. Later when differences in needs and priorities occurred, these had to be resolved amongst the managers themselves. Only as a last resort could they refer to the Director, Consultancy Services.

Outcomes

To what extent were some of the original problems solved by the introduction of the matrix? There are always uncertainties in attributing effects after the event, but a number of improvements can be directly attributed to the separation of management roles and to the introduction of the matrix structure. Other changes would probably have happened anyway, particularly those associated with the acquisition of more general management skills by a group of people whose orientation had previously been primarily technical.

The reorganisation associated with introduction of the matrix was less difficult than expected. There was almost no disruption of ongoing work. Staff had to learn about the changes in managerial role. An immediate effect was that they found they were getting more managerial attention and support in projects. They soon realised they were being more carefully selected for projects, and for a wider range of work.

Improvements in the quality of work and in the service to existing clients were identifiable within the first year. Four main types of influence seemed to be contributing to this. Firstly, project managers were less distracted by responsibilities other than for managing ongoing work. Secondly, the project managers had been appointed according to their knowledge of particular industries which resulted in closer working links with clients. Thirdly, there was an improved allocation of people to projects resulting from choice of people across the whole of the company. This, combined with the introduction of a staff appraisal system, made it possible to achieve a better matching of individuals to work requirements and also brought a greater diversity of skills to projects. Fourthly, there was improved project accounting due to changes in the management information system required by the new project managers. Within a year of the introduction of the matrix these changes resulted in the company's largest improvement in profitability.

Staff also found that their career opportunities had improved. There was more feedback about their own performance from appraisal interviews. They were also being selected to work on a much wider range of projects and were working alongside more people with widely differing skills than previously. Within a year staff turnover had dropped to well below its previous level. (This was also attributable to a drop in the market demand for computer staff.) Within the company there was a noticeable decrease in career frustrations expressed by staff.

Developing measures of accountability

An initial problem was to develop measures of accountability for the *project managers*. Previously all managers had been judged by their staff utilisation. In theory their utilisation should always be 100 per cent, since all their staff were working on projects. In practice there are always reasons why some staff time cannot be charged to clients – mistakes occur in the work, staff have to be trained on the job, original estimates prove wrong, or unforeseen difficulties occur. Under the new system project managers had to explain all write-offs, previously concealed in overall figures of staff utilisation, and quickly introduced controls to keep non-chargeable time to a minimum. Improvements were achieved by more accurate estimating, more sophisticated use of project control techniques, more specific contract agreements, and more clearly defined work standards. As a result there was a reduction in staff (and computer) time written off. Project managers were also responsible for obtaining repeat business from existing clients. This was not an obligation to keep the consultant's 'toe in the door' but rather to ensure that the quality and value of the work invited a continued working relationship. In fact a very high proportion of new work, up to 80 per cent, occurred in this way. Project managers did not always feel that this aspect of their work was fully recognised.

Resource managers were still judged primarily in terms of staff utilisation. Since they were no longer responsible for managing the day-to-day work high utilisation was achieved by ensuring that staff were available and used by project managers. They could achieve this by ensuring that project managers knew of the availability of staff, by bringing in new work of a kind that ensured that existing skills were better utilised, by recruitment and training people to the available work. At times of high utilisation bargaining used to take place between project managers competing for the same individual for ongoing work, or between project managers and resource managers where the latter wanted to use somebody to develop new business rather than have them fully occupied on current work. In most instances these were verbal disputes concerning the best utilisation of the resources of the organisation, taking account not only of current needs but also of future priorities. They were also the disputes most frequently referred for resolution to the Director, Consultancy Services.

One of the skills of resource management is in maintaining an adequate reserve of people. A large reserve meant that project managers always had the staff they required, but this resulted in low utilisation for the resource manager and subsequent pressure from senior management for not covering costs. Resource managers, whose utilisation remained consistently very high, were profitable but under pressure from project managers for never having available staff. Also they were under pressure from their own staff who had little time in which to undertake training. (A more recent refinement of accountability in terms of utilisation has been the addition of 'net contribution'. This takes into consideration both costs and earnings of staff at different levels of the organisation.)

Resource managers were also responsible for selling to new clients. This included both new computer applications and finding new clients for existing applications. They also had longer term responsibility for the R and D aspects of the work. With lead times of up to five years it has proved difficult to develop criteria of accountability that include both the research and the commercial aspects. In any case longer term success is often the result of work and influences in many parts of the company. Whilst some resource managers get greatest satisfaction from the staff management and staff development aspects of their work, others find the greatest challenge lies in anticipating technical developments and realising their market implications.

One consequence of using staff utilisation as the main measure of performance is that there is no necessary requirement for growth in numbers. It is possible for a resource manager to maintain high utilisation and to increase the skills of existing staff, and hence increase profitability, without necessarily adding to the size of his department. Growth has been achieved through the development of overseas companies, partly staffed by people from the matrix, and from better utilisation of existing skills and experience. The present senior management feel that this is the most appropriate use of the company's skills and resources and also the most profitable.

Flexible yet stable

A feature of the matrix has been its stability over the seven years of its operation compared with the previous frequent reorganisations. The resource divisions have provided the continuity of experience both for the development of the work and in meeting people's career needs, whilst the project groups have provided the flexibility necessary to meet the constantly changing demands of the work. It has also proved possible to make adjustments to both the project groupings and the resource divisions to meet changing work or market requirements without having a major effect on either the reporting structure or the definitions of job roles. Adjustments to one resource division or one project grouping do not disturb either the work or the reporting relations in the rest of the matrix. An increasing number of managers and senior consultants have now operated in both the project and the resource functions resulting in a good understanding of the two roles and the ways in which they interrelate.

Several years after the introduction of the matrix the inclusion of the company's computer bureau was attempted. Although sharing a number of common services, the bureau was essentially a separate operation. The rationale was to produce a unified structure for the whole of the UK company and to facilitate the sharing of resources and skills. To be profitable a bureau has to maintain high utilisation of its computers, necessitating three shift working for operators. Much of the operation is of a routine nature. After eighteen months of operation it was recognised that the matrix did not meet the operational requirements of the bureau and it reverted to its separate and mainly hierarchical structure. The attempt demonstrated the extent to which a matrix is suited to situations requiring mixed project teams where work assignments are constantly changing.

Working for two bosses

A frequently expressed doubt about a matrix concerns whether people can effectively work for two bosses. The extent to which this creates difficulties for staff is largely a function of the distinctiveness of the two management roles. If the managers are themselves uncertain of their roles, or if there are competitions for power, then it is likely that staff will experience difficulties. Additionally, some people experience conflict between shorter term project goals and their longer term career aspirations. The main influence on careers occurs in the allocation of people to projects and sometimes people find themselves allocated to work because of its similarity to what they have done already whereas they would have preferred to broaden their experience by being given a new type of work. The same problem can arise when people have been working on the same project for several years and would like a change. At least in the matrix structure these conflicting priorities are reflected in the roles of the project and resource managers so that it is possible for the individual to get managerial support for his position. The most usual outcome of such conflicts of interest is for the individual to be asked to carry on work with which he may be bored but in which he is experienced on the explicit understanding that he will be given new work at the end of an agreed period. Prior to the introduction of the matrix this used to be an important source of discontent and cause for staff leaving since they often felt they had little option if they wanted to broaden their experience.

An unanticipated outcome of the matrix has been the opportunities that it has offered to individualists to find both work content and reporting relationships that suit them best. Any organisation offering consultancy services contains many people who want to work independently or only refer for collaboration as and when they need it. While people's 'back home' resource division has tended to remain fixed, many have experienced considerable freedom in determining the type of assignments and the kind of reporting relationships that suit them best. In this sense the matrix has provided a stable overall structure, but one within which many people have found their own individual niche.

Before the introduction of the matrix there was a constant tendency for increasing complexity in the outside world to be reflected in more complex and more hierarchical reporting relationships. Since the introduction of the matrix reporting lines have on the whole been kept short. Some hierarchical structure is required by both project and resource managers to do their jobs and at various times in the past some have been tempted to create extended reporting lines beneath themselves. However, clearer criteria of accountability combined with senior management discouragement have helped to minimise the number of levels of reporting relationship.

Staff appraisal

The continuous and active selection of people for projects has brought into sharp focus the methods being used to appraise them. While staff wanted an appraisal scheme that provided indicators of their potential and career opportunities, project managers were keen to ensure that they had an accurate picture of people's capabilities before selecting them for new projects. To some extent these two requirements were in conflict. Project managers wanted detailed and sometimes harsh judgements about an individual's capabilities in order to be able to make their decisions. On the other hand, individuals wanted feedback about their performance but there was a marked reluctance to convey this to them in appraisal interviews particularly if performance was less than satisfactory.

While some staff found that there was competition between project managers for their skills and experience, others found that they were not being selected for projects. This left the resource managers with the problem of whether the particular individual had skills or experience that were not recognised, or did not have abilities or experience that met the needs of the company. In the previous structure it was possible for divisional managers to protect staff whose competence might be doubted by others. Individuals were more exposed following the introduction of the matrix since their skills and experience were subject to scrutiny every time they were being considered for work by a different project manager. Very few people left as a result of this kind of pressure but one young management scientist referred to the introduction of the matrix as a form of 'structured insecurity'.

Management style

The introduction of the matrix did not recreate the informal management style that was characteristic of the company during its pioneer days during the 1960s. Shorter reporting lines and clearer criteria of accountability did however help people to know more about what was going on and to be clearer about where they personally stood. There developed closer working relationships amongst people drawn from across the company deriving from the needs of the work. The work became the primary focus of contact between people and the focus of information about what was going on.

The matrix provided a release from the detailed and intrusive controls that characterise a drift towards greater uncertainty whether these occur inside or outside the company. In this respect it provided certain freedoms and the impression from my own experience of working within the company for the three years following the introduction of the matrix was that the company became both more commercial and more confident about tackling areas of uncertainty.

Conclusion

Scicon Group now has a number of subsidiary companies in Europe and elsewhere and consideration has been given to whether a matrix is appropriate to any of them. Some of the overseas companies are not yet much more than a number of separately managed projects with attached marketing and selling functions. In any case many of the staff are members of the back home matrix and are under secondment to one of the overseas companies as part of their assignment to a project. The German company Scicon GmbH is most like Scicon Ltd in terms of size, of the mix of projects and of people. The German senior management however have decided not to adopt a matrix, having expressed the view that 'the line of command in the matrix is not clear'. Perhaps this is an important difference between German and British styles of management, particularly as applied in a high technology industry! John Ockenden, the Managing Director of Scicon Ltd, summarised the present position by saying, 'It is one of the few things in the company that has not changed over the years. We have no ideological commitment to a matrix but it happens to suit our type of people and business very well'.

Summary

Due to a combination of rapid growth and constantly changing demands for computer and management sciences consultancy, Scicon was finding it increasingly difficult to satisfy both customers and the career needs of staff. A change process initiated by a staff attitude survey led to the creation of a matrix structure with separate roles and responsibilities for project and resource managers with distinct but balanced authorities. Project managers became responsible for managing all ongoing work, while resource managers became responsible for selling, marketing, recruiting, staff management, and research and development. The distinction and inter-relationships between the two roles have become refined during the seven years of operation to include clear definitions of responsibilities and accountabilities. During this period modifications have been made to parts of the matrix without disrupting working relations elsewhere.

5   The advertising agency account group: its operation and effectiveness as a matrix group

D. S. FRANKEL

Introduction

One theme that must particularly concern managers is the way the organisations and groups in which we spend our working life are changing. That change is happening, and rapidly, cannot be doubted. When we hear and experience terms like 'open office space', and 'ecological concern' we are dealing with developing organisational situations probably unknown to previous managers. In particular there is increasing concern with what can be thought of as 'the quality of organisational life', and interest in the way individual contributions to the total effort of the company are integrated so that the experience of work is positive and concerned with more than the monthly pay-slip.

Certain trends emerge when we think about the direction in which our organisations are heading. One such trend is towards the project or matrix form of organisation. A matrix group will be concerned far more with the expertise and skills of job members than the formal authority the organisation chart is supposed to grant them. At the individual level, satisfaction from participation in the matrix group is achieved only on condition that the needs and values of the individual members are met. The role of management in relation to such groups is to facilitate their creation and performance, not to set rules and programmed directions that must be followed 'or else'. This allows the matrix members a greater potential to perform work within the group in a manner that is meaningful to their professional identities, and thereby to become committed to the group and the work it is designed to achieve.

In the past, research into this all-important question of work and group structure and design has often suffered from being divorced from real organisations and settings. Instead, research in this area tended to be centred on the artificial laboratory-based 'experiment', in which participants are strangers to each other, participant concern or commitment to the group is seldom generated, and the subjects themselves are usually students, who cannot represent a wider population. We will understand far more about the actual issues, problems, and potentials concerning matrix groups if we look at actual ongoing groups in real organisations and companies performing actual tasks. The question is how? Below, we will consider a particular type of group in order to judge whether it is a 'matrix group' and to see whether the way the group manages its performance and behaviours contributes towards its potential for being successful.

The agency and its account teams

During the period 1972–75 the author carried out a series of investigations into the nature, composition, and belief structures of members of advertising agencies who were directly concerned with the development of advertisements for the accounts or clients serviced by the agency. Each client account is handled by an 'account team'. While the size of this team will vary according to the demands of the client and the resources of the agency, each account team will have a nucleus of four group members belonging to two primary agency functions. These functions are account management and creative management.

Account managers are the link men connecting transactions between the agency and client systems. Each account will consist of an account director (who manages the strategic aspects of the account) and an account executive (who handles the technical or day-to-day running of the account). Through the account managers, the skills and services of the agency are made available to the client. The account manager's task is essentially one of information retrieval (from the client) and transmission (within the agency). Once he has been alerted to a new or revised client need, he attempts to accumulate information about both the factual aspects of the proposed project or campaign and the emotional or affective needs of the client in relation to the project. On the basis of a contact report written by the account manager, an agency meeting is convened to work out a possible advertising strategy. In order to maintain the deadlines associated with advertising preparation, the account manager must monitor and hasten the process of early creative work. The creative team itself will tend not to have much direct contact with the client, and must rely on the account manager to judge whether ideas generated will be received favourably by the client. Once the client has been presented with the preliminary output, the account manager endeavours to obtain his approval. Failure to 'sell' the advertisement requires the account manager to justify, within the agency, his lack of success as well as to indicate how suitable amendments can be made. The cycle of information retrieval and dissemination is repeated to the point where the client accepts unconditionally the final design stage suggested by the agency.

The other function within the agency encompasses those behaviours and skills that use the informational inputs of the account management function as a basis for initiating and designing the theme and nature of the actual advertisement – the creative function. At a minimum, this function consists of an artist who is concerned with visual images, and a copywriter who deals with verbal images. The creative manager has two roles within the agency: one, his role as a member of his functional group (e.g. all copywriters in the agency), and two, his role as a member of the account or project team. A sequencing of the advertising process from the creative perspective is as follows. The creative managers on an account consider the information presented by the account manager about the needs and parameters of the account. An initial concept for the advertisement is developed by the creative managers into a rough artwork stage, and the client's verdict on the applicability of this output is transmitted to the creative managers. If necessary, either an overhaul or a revision of the concept and 'roughs' is performed, and this, once accepted by the client, is converted into artwork or final design stage. Approval of this final presentation signals the transformation of the artwork to a proof stage, ready for insertion into a medium, thus terminating the particular advertising cycle.

The two functions, as described briefly above, form the basis for the account group. It should be noted, however, that we are here referring to an 'ideal–typical' account group upon which variations are possible. While a variety of other functions may contribute to and participate in the development within the agency of an account project (e.g. the media function which seeks to maximise media impact at minimum cost), only one other function was considered by the author to be able to influence and direct the actual inception and nature of the account group processes. This is the 'traffic function'. The role of the 'traffic manager' (although he goes under many names) is to ensure that all written information relevant to the account is centralised, collated, and distributed without time-lag to group members. He must use his persuasive skills and administrative muscle to remind members of deadline pressures, and must exercise control over the quality of the designs and reports issued by group members. In a wider sense, he is the group facilitator – the 'integrator' whose purpose it is to co-ordinate the group decision process, even if his organisational position (or salary) reflects only his administrative accountabilities.

The account group as a matrix group

The question may now be posed: is the account group, as described, a matrix group, and, if so, what properties does it exhibit in the furtherance of its task? The factors supporting a belief that the account group is indeed of matrix design are:

(a) it is composed of agency members, each with a line or functional responsibility to an organisational department, who for the life of the project work inter-functionally together, and whose tasks require that functional areas of skill and expertise are made available to the group within the context of the overall project;

(b) the group is temporary. Should the client fire the agency or should the advertising campaign be fully approved by the client, the group will disband until a new project is initiated. Consequently, the life of the group reflects the life of the project;

(c) the orientation of the group members is professional, and suggestions from members of the group concerning the shape and scope of the account within the limits imposed by the client are considered and evaluated by the group according to the merit of the suggestion, rather than the organisational position of the member. The group recognises expert power above positional power as a determinant of group outcomes;

(d) communications are lateral rather than vertical, i.e. they are conducted according to project rather than according to hierarchy;

(e) the traffic manager acts as facilitator of the matrix process, reconciling divergent viewpoints and encouraging the development of a group perspective in order that the final outcome represents the overall group perspective rather than individual orientations alone. It is often the traffic manager and not the account manager who performs this integrating role, since the former is free of a functional orientation or bias that might invalidate the acceptability of the integrator in the eyes of the other function;

(f) members of the group may belong separately and jointly to other account groups, leading to the existence of overlapping group memberships.

The above points lend credence to the statement that the account group is a matrix group, reflecting the project basis whereby the advertising agency itself structures its work functions. Should this be so, provided we can discriminate between effective and less effective account groups within the same system (agency), we will be able to understand whether such differences in performance are attributable to differences in the way these groups organise themselves in order to accomplish their tasks. The results presented below are based on the empirical and substantive research studies conducted in over fifteen agencies (see Frankel, 1975).

Group effectiveness for account teams

Since the focus of the study was on individual account groups within agencies, when considering how to define 'effectiveness' of the group and its performance attention was centred on the group itself rather than the agency as a whole. This enables one to compare different account groups within the same agency, but not to compare agencies as a whole with each other. In order to establish such a measure of account group effectiveness, a definition specific to advertising agencies rather than' companies in general was designed, viewing effectiveness according to the group's degree of goal achievement. Account group effectiveness is defined in terms of the ability of the account group to produce work that will improve and enhance two separate reputations, both necessary if the agency is to survive. The one reputation is client-based (and is of particular value to the account manager) – does the client like the work, will he increase the advertising budget? etc.; the other is professional-based (and is of particular value to the creative manager) – will the work win an award, will new clients be attracted to the agency because of the work? etc. The more effective group, therefore, will produce work that can be used by the agency to attract both new clients and potential talented staff to the agency to a greater extent than can the work of the less effective account group. Since the competitive nature of the advertising business is such that an 'ineffective' account would be tolerated by neither client nor agency for extended periods, it is more accurate to think of the difference in effectiveness levels of groups to be in terms of 'effective' and 'less effective' rather than 'effective' and 'ineffective'.

Determinants of group effectiveness

The processes used by the group to attain an output acceptable to both the client and the agency were noted to consist of two types of behaviour. The first is *inter-functional behaviours*, whereby the account management and creative management functions work in conjunction with one another. Examples of this are: the artist presenting to the account manager the

designs the latter is to take to the client; and the account director inform-
ing the creative team of client reactions. Inter-functional behaviours have
a dual purpose. Firstly, to allow each function to present to the other
what it is doing so as to secure the other's approval in order that the next
stage of the process be reached. Secondly, to provide each function with
an opportunity to query the other as to its progress and the nature of its
involvement with the project. Within the group, two group mechanisms
are possible whereby inter-functional behaviours are achieved. These relate
to the formal and informal components of group structure and may be
called *co-ordination* and *cohesion* respectively.

Co-ordination here is seen as mechanisms imposed upon the group by
hierarchical means, and which oblige group members either to meet at
a specific time or to exchange written information. Rejection of this
authority by members may be met with punitive measures. Cohesion,
on the other hand, is explained less by hierarchical authority than by the
personal inclination of group members to disseminate information and
relate to each other by choice. No punishment is attached by the agency
to the individuals who do not subscribe to this integration. Among examples of co-ordination are: regular meetings, weekly progress reports,
specific briefing sessions from which no group member may absent himself, etc. Examples of cohesion include: members of group meeting after
work for a drink, group members holding *ad hoc* meetings to discuss an
aspect of the task, and members meeting and communicating on matters
not directly concerned with the account.

While an amount of co-ordination is necessary if the advertising process
is to be controlled, disadvantages attach themselves to an overuse of
formal co-ordinating measures. It may, for example, cause members to be
brought together either before a need for the integrative effort has
surfaced or while one function is still engaged in completing a task which
may suffer if assessment thereof is premature. Cohesion, on the other
hand, is a mechanism related directly to personal needs or desires. This
informal method of achieving inter-functional contact may be both task
and socially oriented. Cohesive acts were demonstrated to enable feedback
and criticism to be forwarded in a more open and supportive way than if
formal mechanisms were used.

In order to quantify these observations two lists of ten items tapping
co-ordinatory measures and cohesive measures respectively were designed
and tested for validity and reliability, and then applied to two account
groups of different levels of effectiveness within the same agency. This
was repeated in three other agencies. In all cases, it was demonstrated that
there were no significant (or apparent) differences in the level of co-ordi-

nation used by the effective and the less effective groups. Statistically significant differences were found in the level of cohesion prevalent in the two groups within each agency. *The more effective account was always the one where members exhibited greater and more developed cohesive tendencies.* This lent support to the view that a non-programmable and highly variable task like preparing an advertising campaign required the group to operate and communicate as an integrated group, and that the more successful groups will be those that engage in informal and relatively unplanned acts of communication, and where members relate to each other on a personal as well as task oriented basis. In other words, effectiveness in a matrix group like the account group is a function not of intensive planning and regular meetings (these did not discriminate between groups) but of the ability of group members to choose for themselves times at which to cohere, and to negotiate between themselves what topics are discussed during these cohesive contacts.

The second type of behaviour is *intra-functional behaviours* – those activities that are carried out by the individual or single function without direct reference or interchange with the other function. The output of this phase is then presented to the group. An example of this is the copywriter who prepares a presentation during the weekend at his home. After a period of observation in a range of agencies, and the design and testing of relevant questionnaires that were then applied to all group members in the effective and less effective accounts, it was noted and confirmed that the individual member's *commitment* to the intra-functional behaviours required for group task completion was influenced by two main factors. These are: the degree to which the member perceives himself to occupy a *central position* within the group (i.e. that he does not feel that he is considered by the group to be external to its main activities), and the degree to which he is able to derive *satisfaction* from working on the account. High levels of both member centrality (taken as a group average) and satisfactions attained (again, taken as a group average) were noted to be associated more with the effective group than the less effective group.

An individual member who perceives himself to occupy a central role within the group is more likely to perform effectively than an individual who perceives himself to occupy a marginal or external position. This is because the former's position enables him to receive incoming informations rather than to spend his time searching for them, and if he needs further information he will more readily be able to make contact with the information holders than the marginally placed individual. In this way, the individual's awareness of the situation, which in the advertising task serves as an input to his performance, is likely to be current, and he is aware that changes in plan can be implemented only after he and other central figures have given assent. An individual with perceptions of low centrality may reduce his contribution in order to reflect this marginality.

The second factor shown to influence the degree to which the group member will commit himself to his role is the degree to which he perceives that working on the account brings him satisfactions and affective benefits. An individual is more likely to remain a member of the group if working with it provides outcomes the individual himself values and where the group's performance will be instrumental in satisfying individual needs. This bears out the point that in a system like the advertising agency, where cause/effect relationships are ambiguous, individuals may use the personal satisfactions derived in their work and from their work relationships as a measure of how deep their commitment to the group should be. Only if the individual believes that in the future he is likely to continue receiving these satisfactions will he be induced to continue investing high effort into his task, and gearing his intra-functional behaviours in order that they both be absorbed and achieve recognition within the group.

Conclusions

The conclusion from an investigation into certain of the determinants of effectiveness in the matrix group within an agency structure was that the group that shows evidence of greater cohesion, member centrality, and affective satisfactions is more likely to have achieved a state of defined effectiveness than a group that shows evidence of these properties only to a lesser degree. In short, the behavioural components of the effective group within the agency context differ from those of a less effective group.

For the manager whose concern it is to facilitate the performance of temporary project-based groups, the above conclusion may suggest how the group's potential may be enhanced. It was noted that the professionally oriented matrix group that was more successful than another was not characterised by formalised communication channels and extensive meetings. No difference in the overall levels of such measures was noted to exist even when the one group was considerably less effective than the other. The factors that do appear to make a difference to the group's effectiveness potential, however, are those that enable group members to choose for themselves times and topics of personal relevance for work activities rather than relying on management to dictate these. In other words, group members should know that they personally are able to influence the direction of group activity. This supports the view expressed at the beginning: that the quality of working life is important, and that the activities of the matrix group should reflect the individual contributions of its members.

Does the above mean that once a matrix group is formed management has no right to influence its performance? Far from it: rather, the opportunity is now open to facilitate the operation of the group, to check that the opportunities for informal cohesive actions are available, that the views of the members are allowed to guide the direction of effort, and that no group member should be excluded from participating in the group according to his/her full potential. Ultimately, it is the challenge of matrix management to integrate the group orientation alongside that of the wider organisational mission the group was originally formed to accomplish. By so doing, the objectives of the group and those of the company itself become unified. This allows for growth, in both the corporate and personal sense.

Summary

Advertising agency account groups tend to be composed of two primary functions – the account managers (who handle client contact) and the creative managers (who prepare the advertising themes and design). This basic group is temporary (reflecting the life of the project), professionally oriented, and must work inter-functionally if an end output is to be achieved. Group processes used are of two kinds – inter-functional behaviours (the two functions operating together) and intra-functional behaviours (carried out without reference to the other function). Working as a matrix, the group that evidences higher cohesion, member centrality, and affective satisfactions is more likely to have achieved a state of defined effectiveness than groups showing these properties only to a lesser degree. This means that the behavioural characteristics of the account group are determinants of the potential effectiveness of the group within a client servicing matrix.

6   Matrix organisation in health services

MAUREEN DIXON

The health service context

Although the term matrix organisation is still generally unfamiliar to those working in the health services, the notion of laterally related teams and working groups has been common for years. There has traditionally been relatively little study of health services and medical organisation – perhaps because it was assumed that the 'caring' professions were not susceptible to the usual organisational considerations – but in the last few years this has all changed. Largely as a result of the 1974 reorganisation of the British National Health Service, there has been a comprehensive attempt to define optimum organisational forms for the delivery of health services, focusing particularly on various models for relating roles laterally rather than vertically.[1] The structures now set up within the NHS are extremely complex and involve the overlapping of functional hierarchical organisation with lateral relationships of various kinds. The multi-disciplinary team is widely used both for management purposes and for the provision of diagnosis and therapy.

For the most part, the professions and disciplines involved in health services have been organised in discrete hierarchies. Many of the professions, nurses and pharmacists for example, have established their own hierarchy with the senior management role being open only to members of the profession and directly accountable to the governing body. This trend towards an increasing number of differentiated hierarchies, so-called functional management, has been marked in the past ten years or so and has inevitably brought with it severe problems of co-ordination.

A major exception to this general form of organisation are the clinical doctors: the consultants in hospitals and the general practitioners. They are not organised hierarchically but are individually in contract with a health authority. They are not subject to managerial controls and are in a peer relationship with each other. So in addition to the need to link the various management hierarchies there is the need to involve clinical medical staff in multi-disciplinary groups and teams. Depending on the purpose of the group or team, the doctor is involved as an individual clinician or as a representative of a group of medical staff. But whichever of these two roles the doctor occupies, he is subject to different pressures and sanctions from those who have managers outside the team.[2]

So the tendency to create differentiation on the one hand and the recognition of the need for co-ordinated activity on the other have led to the development of a complex matrix organisation in health services. Some of the lateral groupings are permanent, others are *ad hoc* and disband when their task has been completed; some are concerned primarily with the development of plans and policies, others have the direct provision of services as their task; some involve only health service professionals, others extend into other, non-health agencies to include staff of social services and education departments. So, for example, there is the District Management Team comprising the district administrator, district nursing officer, district community physician and district finance officer. This team carries responsibility for both management and planning functions in the health district and the individual members are directly accountable to the governing body, the Area Health Authority. The so-called Health Care Planning Teams are another example of a grouping which cuts across different hierarchies. These teams are set up to identify health care needs for a particular group in the community (the elderly or the mentally ill, for example) and to develop plans for services for those groups. They may be temporary or permanent and often include health service staff and staff from other agencies. Yet another variant of matrix organisation is represented in the numerous teams of actual providers of care such as doctors, nurses, social workers, psychologists and so on. This latter type of team is the one which is analysed in more detail in this chapter to illustrate the complex relationships which arise.

The reasons for matrix organisation

All the variations appear to have developed naturally to meet different aspects of the need for co-ordination. The first of these is the rather obvious one that the provision of health services is, by its nature, a multidisciplinary activity. Even the apparently individual practice of the surgeon's skill in the operating theatre is dependent on the co-ordination and programming of a whole series of other skills and activities. In the modern National Health Service there are few if any activities which can be carried out by one profession or discipline in isolation.

A second need to which the matrix form of organisation responds is the bringing together of different professions and disciplines in a situation where they can learn from, and be affected by, each other's skill and knowledge. The total potential of the team is thus greater than the sum of its individual parts. It reduces the likelihood of duplication (or omission) of activities, it provides the customer or client with a wider range of professional help and it gives its members a point of identification in the actual provision of the service. Perhaps, in the health service context at least, the distillation of the reasons for the emergence of matrix organisation is the developing awareness of the complexity of individual and social problems. Single dimension solutions have been found largely ineffective – the hospital may 'cure' an elderly person suffering from some illness but cannot tackle the problem of the poor home environment which precipitated it. Hence team structures have developed in which different skills and expertise are combined in an effort to match the complexity of need and demand.

However, it is obvious that matrix organisation has other, less prosaic origins. For many contemporary writers it is seen as the answer to the problem, which Toffler, tongue-in-cheek, sees as: 'man as a helpless cog in some vast organisational machine . . . each man is frozen into a narrow, unchanging niche in a rabbit-warren bureaucracy. The walls of this niche squeeze the individuality out of him, smash his personality, and compel him, in effect, to conform or die' (Toffler, 1970). The matrix form of organisation, emphasising lateral rather than vertical relationships, is widely viewed as the solution because of its tendency to increase discretion and participation in decision making and to encourage collaboration between a wider range of people. The egalitarian nature of the lateral work group is constantly stressed (Argyris, 1967). These kinds of reason often seem to be the major motivation for the development of matrix forms of organisation in health services. So, for example, in the clinical situation the development of the multidisciplinary team is as much a reflection of the wish to move away from the traditional view of the omnipotent doctor as it is a response to functional demands.

The main variation in the models proposed is the extent to which the lateral relationships are emphasised and seen as legitimate, at the expense of the vertical (Kingdon, 1973, p.26). Bennis's model perhaps goes furthest in eliminating hierarchy; he predicts the end of bureaucracy in the next twenty to twenty-five years, its successor being the type of organisation in which executives and managers will be differentiated, not vertically, according to rank and role, but flexibly and functionally, according to skill and professional training.[3] Other writers, less confident of the demise of hierarchy, nevertheless see the team, autonomous work group or whatever the title, as essentially a self-regulating group of people deciding on their own internal relationships and action (e.g. Vickers, 1973).

Difficulties in implementing matrix organisation

The matrix model in which vertical, hierarchical relationships are minimised has the same appeal in health services as in commercial and industrial organisations. But practical experience in developing and implementing team structures in the NHS has also demonstrated that it is not quite so simple. The prevailing ideology may support equality, self-determination and so on but the realities of organisational life often contradict the ideology. Take, for example, a team set up to provide psychiatric diagnosis and treatment for children and comprising a consultant psychiatrist, an educational psychologist, a psychiatric social worker and a child psychotherapist. It is assumed that the team is an essentially egalitarian form of organisation. Freed from the restraints of a hierarchical management relationship, team members relate to each other as equals. The voice of each member carries equal weight and decisions are arrived at by consensus, each member respecting the special skills and knowledge of the others. So, it is argued, individual freedom is vastly increased, the only limitation on independence being the need to accommodate to the wishes and freedom of the other team members. The team is essentially non-bureaucratic because it cuts across the traditional vertical lines of authority and accountability and it is essentially participative and democratic because of the lack of a manager or clear authority figure. In the health service setting this model has particular appeal for the non-medical professions because it appears to break away from the traditional situation of medical dominance (Gray, 1974).

But the chimeric quality of these advantages is shown as soon as there is a difference of view within the team. Suppose the consultant psychiatrist has a different idea of the best treatment for a child from, say, the educational psychologist; or the psychiatric social worker considers a whole family should be involved in the child's treatment whereas the psychotherapist feels intensive work with the child alone would be more beneficial – how can such differences be resolved and what sanctions can the team bring to bear on its members to ensure they follow the agreed action, if agreement can be achieved? In practice, the real relationships tend only to emerge when some kind of crisis is on hand. Then there is generally a return to more traditional distinctions between the different professions

85

and to emphasis on individual, rather than team, accountability. So to see the team as an easily self-regulating group is to underestimate seriously the tensions between the demands of the team and the demands of the larger system in which the team is placed.

In health services these tensions arise from a number of sources, some of which are evident in other public service and industrial organisations. The most obvious source of tension is the fact that team members are also members of different professions. To a greater or lesser degree these professions have developed standards and skills which are vitally important to their members and are the basis of their expertise. So the individual can be faced with the dilemma of choosing a line of action which is consistent with the ethic of their professional community or following the team's policy. These professional pressures are particularly obvious when the individual team member is in a hierarchical situation with a professional manager outside the team.

Another factor militating against the easy working of the matrix structure is the special and traditional role carried by the doctor. Characteristically in the British NHS the consultant (or GP) has accepted individual accountability for the patients under his care and has thus been able to exert some authority on matters of diagnosis and treatment over others who are not his managerial subordinates. This position is embodied in medico-legal decisions over the years. But the multidisciplinary team is apparently based on opposite assumptions – the consultant is no longer personally accountable, so he carries no special authority with regard to other members of the team. Accountability can be carried by the team as a whole or, if agreed, by any individual member. This model causes all sorts of practical problems for team members since they have generally been trained and educated to the more traditional medical model and because the expectations in the outside world, legal and other, are based on the idea of the individually accountable doctor (Friedson, 1970).

Mention has already been made of the multidimensional nature of the problems which the health services, like others, are having to tackle. One response to this has been to develop organisational links between totally separate agencies, to extend the matrix across organisational boundaries. For example, in the child psychiatry team the consultant psychiatrist is in contract with a health authority in the NHS, the educational psychologist is employed by a local education authority, the psychiatric social worker by a local authority social services department and the child psychotherapist by an area health authority. But, of course, to the extent that the number of agencies involved is increased, so is the possibility of conflict and difference of view within the team. The employing authorities inevitably wish to retain some kind of link with their employees and are not content to set up the roles in a way which does not provide for some sort of accountability.

So the team within a matrix in health services frequently produces not just a dual-influence but a multi-influence situation for its members, responding as they do to pressures from the team, their employing authority and their profession. As if this were not difficult enough, there is a further very real source of tension in the remuneration and status systems. Because of the different employing agencies and the fact that the systems of remuneration and salary scales are separately worked out for each professional group, there can be great differences in levels of pay within a team. However Herzbergian (Herzberg, 1966) one's views of the importance of monetary rewards for work, it is clear that the differences in pay do lead to feelings of difference in status and power within the team. It is difficult for a newly qualified, relatively lowly paid social worker to have the same impact on the team as a consultant of twenty years' standing.

Clarifying responsibilities

Seminars or conferences of health service professionals today are more likely to be concerned with team relationships and lateral linkages with other organisations than, as was the case ten years ago, with managerial skills and how to develop them. Lateral teams, in one form or another, are a constant feature in the working life of most staff in the health services. The sort of questions under discussion are likely to be: how to achieve co-ordination of the team; how to distinguish between individual managerial responsibility and team responsibility; how to arrive at decisions in the team; how to reconcile the demands of an external manager with the demands of the team; and how to decide which team member has the most appropriate skills to deal with a particular issue. These are highly practical problems, all related to accountability. The tension for the members of the organisation arises from the fact that they will be held accountable, but that it is often ideologically unacceptable to specify what this accountability means in terms of authority. If specification is attempted, it is often in oversimplified terms with inadequate concepts. There is a tendency to work with only two conceptual models – the management hierarchy or the peer group (cf. Galbraith, 1973, p. 48). The choice is thus seen to be between maximum control via hierarchical authority and group control by peer influence.

As long as the conceptual armoury is so weak, it is impossible even to explicate the differences in team members' perceptions of their relationship. For example, two members of a team may convey concern about the lack of a clear leader in the team, but to one a leader may simply be an equal who co-ordinates the group, whereas to the other a leader is an accountable director of some kind. It is simply not possible to understand or attempt to solve problems of relationships within the team in terms of undefined concepts such as shared responsibility and leadership. In research in the NHS it has been found necessary to develop a range of concepts of working relationships which does justice to the complexity of these matrix situations.[4] So, for example, the role of co-ordinator has been defined and has obvious applications where it is felt necessary to establish one person with the function of co-ordinating the work of a number of others in some particular field and where a managerial, supervisory or staff relationship is inappropriate – for example, in the lateral team. Similarly, the concepts of co-management, outposting and secondment provide alternative ways of relating an individual from a management hierarchy to a lateral grouping cutting across hierarchies. Using such concepts, it becomes possible to unravel different perceptions of the existing situation and to consider possible future forms of organisation.

Another factor which is important in clarifying responsibilities in the matrix is that organisations are fundamentally about accountability. When an employment system is established, usually by an association of some kind (Brown, 1971, pp. 48–9), it is intended to enable work to be carried out to achieve the objectives of the association. The individual who takes on a role in the system enters into a contract with the association, not to exercise authority, but to be accountable for duties attached to the role. Exactly the same is true if the individual is employed in a non-hierarchical role to provide personalised services, as in the case of the consultant in the NHS. Accountability is the organisational chicken, not the egg.

The point may seem too elementary to be worth emphasis but the previous discussion has shown that popular notions about matrix organisation are construed in terms of authority rather than accountability. The lateral work groups are created to produce a particular authority setting, ignoring the fact that there are many, and often conflicting assumptions about the team's accountability. An individual is then understandably disconcerted when he is expected to be accountable for something he regards as a team responsibility. Conversely, if there is an attempt to hold the whole team accountable for a situation, the members are forced to decide *post hoc* which of them is accountable for what aspects. The result is that it becomes extremely difficult for the employing association to hold anybody accountable for anything. So the first question in designing a matrix organisation must be: what kind of accountability system do we wish to create?

Research findings suggest that there is another dimension of organisational design that is crucial in matrix organisation, that is, the characteristic levels of work of the different members of a lateral team. Fashionable as it is to espouse ideas of organisational equality, the fact is that work itself is not equal or the same. Work differs in respect of the state of mind of the person doing the work, or some characteristic of the person, so we speak of boring work, challenging work, skilled work, and so on. But it also differs, in itself, in the decisions it involves and the discretion exercised in making those decisions. Research work in health and social services has identified a series of discrete, qualitatively different strata of work to be carried out in organisations, which may or may not coincide with the formal statement of hierarchy and grades.[5] It seems that the work to be done falls into a hierarchy of discrete strata in which the range of objectives to be achieved and environmental factors to be taken into account successively broaden and change in quality. The work at higher strata is felt to be more responsible but significant differences of responsibility are also felt to arise within strata. At least five such strata can be identified in terms of the decisions and discretion expected in each. These strata form a natural basis for delegating work and thus for constructing an effective chain of managerial levels; even where managerial relationships are excluded for other reasons, as in the lateral team, the question of the kind of work to be done by whom still remains.

Going back to the example of the child psychiatry team, the work strata model provides a useful insight into the practical alternatives. If some members of the team are working at a higher stratum than others, it would be artificial, and felt to be so, to suggest that each member of the team has equal authority or power. The forces in the situation would push the members working at the higher level towards taking on some authority over the others. Since this is felt to be inconsistent with the egalitarian ethos of the team, both sides of the relationship feel discomforted. Equally, it would predictably be difficult for a team member working at the same stratum as the others to take on a relationship to them which involved strong authority and sanctions.

Summary

Matrix forms of organisation are familiar in health services but have only recently been the subject of much study. The National Health Service is the classic case for matrix organisation, since it is highly differentiated functionally but must co-ordinate its planning and provision of services.

The lateral team has the advantage of bringing together different professional skills and agencies but also typically brings its own problems of relationships. These problems concern both the internal relationships within the team and individual relationships to the larger system.

In the course of research in the NHS it has been found necessary to develop a range of concepts which adequately describe the many and complex working relationships which arise in teams. The basis for clarification is accountability and the intrinsic levels of work which are to be carried out.

Notes

[1]   Department of Health and Social Security, 'Management Arrangements for the Reorganised National Health Service', HMSO (1972).

[2]   For a fuller analysis of the basis for the independence of the clinician in the British NHS, see 'Working Papers on the Reorganisation of the National Health Service – Revised October 1973', Health Services Organisation Research Unit, Brunel University.

[3]   Bennis, W. quoted in Toffler (1970).

[4]   The process of developing these concepts is referred to in Chapter 10. The main definitions resulting from this work are set out in Rowbottom et al. (1973), Appendix A, in Jaques (1976), Chapter 17, and in working documents available from the Brunel Institute of Organisation and Social Studies.

[5]   The ideas in this section are taken from a model developed by Ralph Rowbottom and David Billis, Brunel University. It is interesting that the quality which differs from one work stratum to another in this model is close to the 'diversity' and uncertainty of the basic task' identified by other writers as being the best basis for organisation (Rowbottom and Billis, 1977).

7 Local authority social services departments: examples of matrix organisation

ANTHEA M. HEY

Introduction

In the last seven years local authorities' social welfare services have been subjected to a great many changes and demands, all of which have carried organisational implications. In order to present a clear picture of the matrix patterns which have developed in some authorities it is necessary: first of all to set these departments in a general historical and developmental context; secondly, to be clear about the kinds of services they are expected to provide and other work they have to engage in to support their services; thirdly, to understand some of the organisational assumptions which are common amongst them, for it is from these assumptions that matrix patterns emerge. Emerge is probably the appropriate word to use in this context because the term matrix is not one which is commonly used. It is the rare department which has explicitly said 'we are going to use matrix organisation or management in order to get effective services delivered'.

History and development

Before 1971 there were several local authority departments providing social welfare services. Some of these, notably children's departments, had grown in size quite dramatically in the middle 1960s as a result of growing legislation. Increase in size had in fact already brought into the open concern with organisational matters.[1]

In 1971 these separate departments were brought together to form one department.[2] These new departments were obviously larger, not simply as a result of merging former departments but because the purpose of the merger was to establish a strong department which could make a more comprehensive response to social need. In particular, emphasis had been laid on research and planning, and staff were recruited for this as well as for the increased need for programming and personnel work which the larger department itself gave rise to.

In 1974 most local authorities outside Greater London were reorganised again, and a significant number of these became larger as more local authorities were merged. At the same time social workers previously employed by the National Health Service were transferred to the employment of local authorities. The expectations on the revamped National Health Service, in particular for more systematic planning and resource allocation, in turn re-emphasised the need for research and planning activities within social services departments if the two agencies were to achieve a more rational service for those in need (BIOSS, 1976). This approach was matched by the 'corporate management' movement explicit in local government reorganisation. The recognition of interdependence between agencies seems to need increased personnel to deal systematically with their interactions.

So although increased size is commented on frequently, there is critical reflection on the increased number of senior posts, particularly when these have been filled from the ranks of experienced practitioners. Mostly the commentators fail to understand the increased demands on the department as a whole, in particular the major operational and strategic planning work which is required. This failure is understandable when at the same time new legislation catering, for example, for the chronically sick and disabled, and for children in trouble, has demanded increased ranges of services without the necessary funds being made available for either capital or staff resources. Demands have not, however, just been for increased ranges of services, but for improved quality of service all round to match the rise in the general higher expectations in standards of living.

These demands have led to employment of more professional and occupational groups, for example, planners, operational researchers on the support side and community workers and occupational therapists in direct work with clientele. At the same time there is developing amongst existing disciplines a greater degree of specialisation, particularly perhaps in treatment matters, for example, family therapists and group workers. Increasingly there are career progression opportunities for those staff who wish to continue as practitioners.

In the report of the Committee (Home Office et al., 1968) which recommended the bringing together of the various social welfare services, a great deal of emphasis was placed on the need to set the client, or potential client, in his social context. Comprehensive social work services were to be provided from area offices situated where they were most accessible, the one door for the individual or family in distress. Such a recommendation has been responded to by increasing decentralisation of services and accountability. This trend can be linked with the corporate approach referred to earlier and to in-vogue models of social work intervention (Pincus and Minahan, 1973; Goldstein, 1973) based on systems theory. All of these variables have tended to reinforce the wish to be more comprehensive in the way needs are being met and scarce or expensive resources are being utilised.

Lastly, but not least, social services departments have been faced with increased demands for participation in decision making processes by both clients and employees. There have been many more occasions of industrial dispute or political action involving social workers in the last few years than at any time previously and the awareness of this has inevitable effects on organisational life. There are various experiments in consultation under way but, more particularly, many departments have sought to provide opportunities for staff who are in contact with clients to share their experiences and contribute their ideas to planning through the formation of working parties and special interest groups whose deliberations reach senior management (Algie, 1970).

The functions of social services departments

Within this rather turbulent environment, what work are these departments expected to engage in? Social welfare is a very general phrase, what more precisely does it involve? The services are very variable, from the provision of bath mats and raised lavatory seats for the elderly infirm on the one hand, to intensive psychotherapy with marital problems on the other. For simplicity it is possible to group the operational activities in general terms into four categories as follows (Social Services Organisation Research Unit, 1974).

At the community level	co-ordinating and assisting voluntary welfare effort;
	stimulating self-help groups amongst those in need;
	registering and inspecting voluntary activity;
	creating public knowledge of services and entitlement to them;
	mass screening for social distress.

With individuals and families providing:

Basic social work	assessing need, planning response;
	giving information and advice;
	monitoring and supervision of clients at risk or subject of court orders;
	counselling in social and interpersonal problems;
	co-ordinating provision of other services.
Basic services	accommodation, food, heat, domestic equipment and social/recreational opportunities;
Special services	occupational training and sheltered employment;
	communications and mobility training;
	aids and adaptations for the physically and mentally handicapped.

In order to sustain and co-ordinate these activities SSDs have a need to engage in: research and planning; personnel and training work; financial work; logistical work, e.g. buildings, supplies and maintenance; secretarial work and public relations work. Depending on local situations they will employ specialists in these functions themselves or utilise those employed by the local authority. In many larger authorities such specialists will be employed at both local authority and departmental levels.

Organisational assumptions

Our researches have shown (SSORU, 1974) that the basic texture of departmental structure is hierarchical, that is, that there is a structure of successive managerial roles, perhaps an optimum of four levels.

There are probably four possible bases for organising social services departments. The four are client (children, the elderly, the mentally handicapped, etc.), function (fieldwork, residential care, domiciliary care, etc.), place (geographical divisions, area or patch), and discipline (social worker, occupational therapist, home help, etc.).

Client base

The Seebohm Report (Home Office et al., 1968) in effect had said that organisation by client group was to be eschewed, and, by and large, that edict was followed. However, more recently some departments have returned to the client group focus, albeit each wider in their concerns and services than any of the pre-1971 separate departments.[3]

Functional base

More emphasis had been placed in 1971 on the social worker operating in a limited geographical territory which he might come to know intimately. However, social work (so called fieldwork) is by no means the only work the department has to carry out. In actuality it is the least costly service in direct financial terms. The most expensive services are residential, day and domiciliary services (e.g. home helps). However, given the Seebohm emphasis on area offices and social workers, it is perhaps natural for those who had to organise the new departments to think first of organising them all in one fieldwork division and thereafter to assign the other work to complementary divisions. In any case, the largest of the former separate departments had mostly organised themselves in this way, i.e. according to some grouping by function.

Geographical base

In 1971, however, one department took geography as its main basis of organisation;[4] although initially all functions were not delegated to division or area there was a stated intention to work towards this end, an end which was realised before the further reorganisation of 1974. This occasion was taken by some authorities to remodel their social services departments and many more decided to take geography rather than function as their main base. As one might expect, this step was taken mostly by the non-metropolitan county authorities, who have large tracts of land to serve and dispersed populations and perhaps substantial urban pockets, market towns or even former county boroughs which are natural social centres.

Discipline base

At this point we are not aware of any department whose prime base of organisation is by discipline, though there are many who have subsections organised by discipline groups. In case some would argue that function and kind of worker would necessarily be congruent, this is generally only the case at present with fieldwork divisions which are largely or wholly made up of social workers. Residential and day care divisions may need to call on a wide range of disciplines to cater for their clients' needs. Domiciliary services divisions, where they exist, frequently include occupational therapists (amongst other disciplines), but the therapists may work also with clients in their own homes and within residential and day centres. In practice there is a good deal of confusion about which professional and occupational groups should be tackling which kind of work. For the present the activities expected of various roles vary considerably from department to department. Until some consensus is established disciplines do not offer a reasonable base for prime departmental organisation, though interestingly it is from the existence of numerous disciplines that later examples of matrix organisation will be taken.

Consequences

For the most part then social services departments have chosen to organise on the basis of function (that is, fieldwork, residential and day care services, domiciliary services) or by place (geographical division or area). It may be apparent, however, that to pose the issue of departmental structure in terms of such straight choices is to oversimplify, for what emerges from detailed consideration of alternatives is that each requires compensating arrangements if due regard is to be paid to a number of co-existing needs (SSORU, 1974).

If, at the first layer, operational work is split according to place, then at some lower point it is further split by function, and *vice versa*. Even if organisation according to disciplines is not a possibility for a prime base there is, nonetheless, a continuing need to take account of the needs of each discipline. Certain groups, like home help organisers, occupational therapists and administrative staff, dispersed as they may be throughout a large county area, will need a senior member of their own discipline attending to their special requirements, recruitment, training and career structure and advising those planning new developments how this discipline can best contribute.

It is when the exploration moves to these compensating arrangements that examples of matrix organisation actually occur. However, before moving to consider such examples in detail it is relevant to note that whether or not matrix patterns emerge in social services departments depends on two further key considerations. The first of these relates to the ambiguities of functions and the boundaries between roles. If it is assumed that social services departments are concerned just with a single field of activity, then it will probably be assumed that one manager, particularly one who is a social worker by training and experience, will be able to manage all the activities in a given area and that it is perfectly possible for him to be accountable for the work not only of social workers, but of occupational therapists, home help organisers, community workers and administrators too. This assumption will almost certainly be reinforced in the local authority setting by the second assumption that any more senior member of staff interacting with a more junior one is a manager, i.e. can give authoritative instructions and can make (and act on his) judgement of the junior's performance. The requisiteness of this is not always considered and there is, generally, lack of understanding of various other role relationships (Jaques, 1976). When these assumptions are made the organisation chart will be drawn as in Fig. 7.1. However, when such arrangements are questioned it usually emerges that the non-social worker staff have ongoing links with senior staff of their own discipline. In many cases the organisation chart acknowledges these links by dotted line connections as shown in Fig. 7.2.

If this figure is multiplied several times, as is often the case, as departments have anything from four to twenty such units, then the organisational complexity begins to be more apparent. It is these dotted line connections which bear examination as possible examples of matrix organisation.

Matrix organisation

Introduction

Analysis has shown that the dotted lines carry different meanings in different situations and, it should be stressed, even within the same discipline. Indeed, it will be shown that in many situations there are real choices as to which meaning to adopt explicitly as the most requisite. When the meaning of the dotted lines is not spelt out they are open to alternative interpretations which may create dysfunctional tensions. For example, if an area officer, as in Fig. 7.1, assumes he is accountable but a senior home help organiser meets regularly with the home help organisers, and her comments are taken to be authoritative, the home help organisers may start to experience conflict in acting upon perhaps contradictory expectations from the two more senior staff members. They in turn may start to feel that they are no longer able to guide the work in ways which are consistent with their accountability. As far as the subordinates are concerned, in each situation they find themselves subject to influence from more than one senior staff member. It is this duality which makes for the overlay or matrix pattern of organisation. There are at present five significantly different situations identified for which it is possible to spell out the precise duties and authority of the senior staff members. These different role sets will be considered in turn. They are:

98

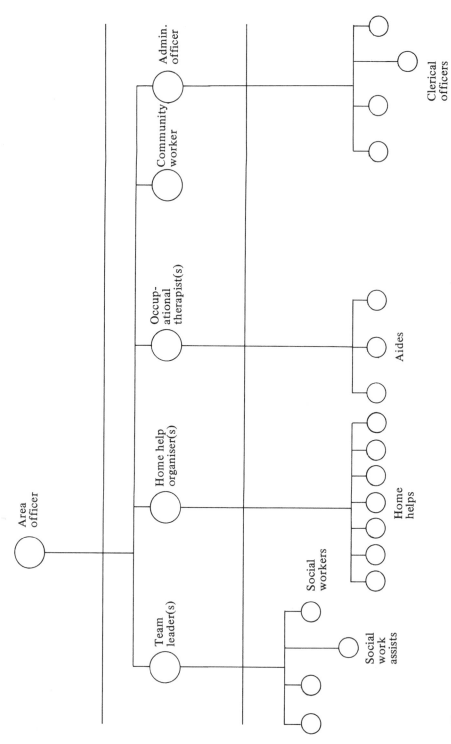

Fig. 7.1   Social services area: simple organisation chart

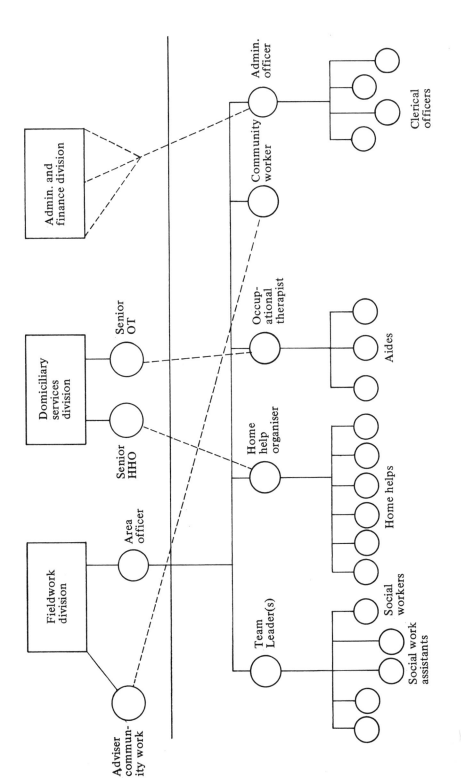

Fig. 7.2 Social services area: complex organisation chart

99

(a)   co-ordinating;
(b)   functional monitoring and co-ordinating;
(c)   attachment with outposting;
(d)   attachment with co-management;
(e)   secondment.[5]

Co-ordinating

Reference has been made earlier to the way in which SSDs have responded to the pressures to provide opportunities for employee participation in decision making by setting up working parties or special interest groups who feed in their deliberations to senior management. They may operate with a specific brief and with a chairman or convenor and individual membership decided by senior management. This mechanism is usually called a working party. The group may meet frequently but only for a small proportion of the overall working week. Work will arise for those appointed in preparing themselves for the actual meetings, for instance, by needing to collect data or views from other people. However, the normal lines of accountability for what they do will not be interfered with. Their managers will need to ensure that they are performing adequately in this task as in any other, and give them any necessary help. The chairman or convenor will carry no more than co-ordinating authority, that is, he will be expected to negotiate the tasks to be pursued and who will do them, to keep the members informed of requirements or progress, to ensure that an adequate report of the deliberations is prepared. In this situation consensus in recommendations need not be required – senior management will decide at the end of the day what action, if any, to implement. However, if disagreements arise about the task itself, then the co-ordinator has no authority to make an over-riding decision and he must report the problem back to those who have established the group.

Special interest groups may be less structured – open to any who wish to join them and appoint or elect their own officers. No more than co-ordinating authority will arise, however, and managers of any staff attending will doubtless want to be sure that the meetings have a useful purpose in relation to the time expended in them. Either of such groups may include staff from different disciplines and from different levels and locations and sections of the organisation and staff may belong to several such groups.

There is a general assumption about these groupings that they are a good thing. They first and foremost provide opportunities for staff participation, but they are often also assumed to be creative and developmental both for the individuals who participate and, thereby, for the organisation as a whole. It is assumed that increasing co-lateral interaction in these ways helps staff to increase understanding about needs and organisation as a whole, leads to improvement in present practices and may be the seed-bed of novel responses in future service provision.[6]

Functional monitoring and co-ordinating

This second kind of inter-relationship also should not interfere with normal managerial relationships. In any of the dual-influence situations shown on Fig. 7.2 it may be assumed in a particular department that the area officer is perfectly able to be accountable for and manage the work of staff of different disciplines, provided that there is some general policy and procedural instruction on how work should be done. It may well be the job of the senior home help organiser or administrative officer at head-quarters to secure that such policies and procedures are developed for their particular function or discipline. They may meet quite regularly with staff of their discipline, based in areas, in order to develop a picture of ongoing operations and to identify any needs for change in policy and practice. They may be regularly consulted for advice or for interpretations of policy. However, they will not have authority to decide new policies unilaterally. Any changes will have to be negotiated with local managers and will probably need sanction from the departmental management group. Sustained disagreements or difficulties in operating sanctioned policies will have to be referred back to the management group for resolution.

Within the examples given in Fig. 7.2 this relationship would commonly be chosen for the relation between senior and junior administrative staff and their area officers and sometimes between senior home help organisers and home help organisers in the field, but not invariably so, as it may be felt that even stronger influence needs to rest with the senior staff of the disciplines involved.

Attachment with outposting

In this situation it would be more accurate to reverse the diagram connections and to show the continuous line from the senior staff member to the area-based member of his discipline and show a dotted line from the area manager. For, in fact, the actual assumption is that of belonging to administration or belonging to home help or occupational therapy. An inference is that anyone who is not a member will not wholly understand its ways of working, its knowledge base, its technology, its ethics and culture as well as the finer points of its procedures and practices. It will be argued that only a more senior member of that discipline can really judge what sort of work a more junior member of the discipline is suited to carry out, to see that he gets it, to help him with his work problems where they arise, to appraise his actual performance and to judge what adjustments or action to promote his development are called for in consequence. In other words, although the junior is located in a place of work at some distance, the managerial relationship is retained by the senior member of his own discipline. In the examples given this might be decided for either home help organisers or occupational therapists.

The area officer in this case will necessarily be concerned with co-ordinating the activities of this attached worker with those of his own staff; both in terms of broad programmes and priorities, and because the nature of social services work may call for several workers to be involved with the same client family, their activities in these cases will need to be co-ordinated too. The area officer will also necessarily be concerned with monitoring the adherence of the attached worker to a local policy and practice, including routine issues of time keeping, leave, and security. He will necessarily require authority to discuss deviations with the worker and to report to the worker's manager continuing and serious deviations.

Attachment with co-management

In turn attachment is used commonly in another situation in which the elements of the managerial relationship are truly shared by the senior officer of the discipline and the area officer – who act as co-managers. This relationship is requisite if there is an ongoing need to balance functional know-how with day-to-day operational judgements. Diagrammatically this could best be indicated by connecting the junior worker by continuous lines to both the area officer and the senior member of his own discipline.

To choose explicitly to allocate an individual two managers is a weighty decision because of its obvious potential for conflict. It is necessary, therefore, to spell out the precise duties of each manager [7] and in particular to note that such an arrangement relies on the existence of a 'crossover' manager who can set or approve policies which are binding on both co-managers and who can adjudicate on any unresolved issue which divides them. As in the last model this one could also be considered requisite for organising home help, but perhaps particularly occupational therapy services where there is a clear professional distinction – a well-established training and high technical content which puts the activity beyond judgement by anyone who is not a member of that discipline.

Secondment

In social services this fifth model of structuring work is so far rarely adopted, though there have been some examples of time-limited projects to which staff from various disciplines and/or sections have been allocated full time. The project will have been under the management of a senior member of staff and for all day-to-day purposes the relationship with the original manager will have been severed. It is assumed in this case that the new manager can adequately encompass all the activities to be able to sustain a meaningful managerial role. The original manager may properly enquire how his worker is getting along, for he will wish to decide how best to deploy him when he returns.

Conclusions

Any one of these five models may be seen as promoting a matrix (overlay) pattern. They arise because of the need to get conjoint work by different disciplines in different locations. As for the specific roles in Fig. 7.2, as indicated any staff may be involved in co-ordinated groups of various kinds in an ongoing way; similarly it is possible that SSDs will come to make more use of time-limited project groups. If they do, a first decision will be whether the project 'leader' can adequately manage all members of the group. If he can, then they will be regarded as seconded. However, just as it is indicated that there are real choices to be made in the permanent structuring of staff between functional monitoring and co-ordinating, through attachment with outposting or attachment with co-management, so will the same decisions need to be made as to the requisiteness of relationships in time-limited projects. Indeed it is possible to argue that if a social services department has chosen geography as its prime base of organisation, it is implicitly, if not explicitly, regarding areas as projects. That is to say that the needs of district communities will need to be responded to comprehensively; that in order to do this it will be necessary to bring to bear the skill of numerous disciplines. It is probable that not all the disciplines can be requisitely managed by the local area officer, who will not share the expertise of all those concerned. He will, however, need to be able to organise those activities. Whilst there are real choices as indicated, for the time being it could be argued that:

(a) administrative staff can be managed adequately by the area officer, provided that there is a functional monitoring and co-ordinating support;

(b) community workers can also be adequately managed by the area officer with similar central policy support and the benefit of ongoing special interest groups from which the authority-wide policy can be co-ordinated;

(c) home help organisers and occupational therapists can be attached with co-management or outposting but cannot be wholly managed by the area officer who does not belong to those disciplines.

Summary

This chapter analysed the history and context and work of social services departments and examined the choices available as to prime bases of organisation by client, function, place or discipline. It noted that function and place are currently favoured bases but that there is some evidence of a shift towards client bases. In any event, it made a strong case for additional formal structural arrangements which can ensure that the work of the various disciplines involved is properly deployed and developed. It is in the attempts to achieve this at area office level that the examples of matrix organisation are found.

Five types of matrix relationships were explored – co-ordinating, functional monitoring and co-ordinating, attachment with outposting, attachment with co-management and secondment. These were demonstrated as real choices but a tentative generalisation of which pattern was likely to be most requisite was offered for the administrative, community work, home help and occupational therapy staff involved. The point was made that whilst in social services departments arrangements are usually ongoing, the same considerations need to be employed in building time-limited project groups.

Notes

[1]   Indeed, the Social Services Organisation Research Unit at Brunel University, of which the author is a member, was first established in 1969 as a result of earlier approaches (1967) to provide training for current managers in children's departments, rather than in anticipation of the major changes already on the horizon. However, the Unit has had a unique opportunity to collaborate in exploring the needs which these departments have been defining and the organisational responses they have made. The

observations in this chapter are drawn from this experience and the author acknowledges the contribution many colleagues have made to her own thinking.

[2]   Education welfare services remain, in most local authorities, the responsibility of the education committee and not the social services committee.

[3]   An example would be Oldham MDC which from April 1976 reorganised in three client groups all those services which were not decentralised to area offices: in practice, residential and day centres, various specialist social work and special services.

[4]   East Sussex County Council. For further explication see SSORU (1974). Although superficially no different at first from many others in organising field social workers in decentralised areas, it was their intent and action in decentralising residential, day and domiciliary services which demonstrated the difference.

[5]   These relationships, or earlier versions of them, were first teased out in work in the Glacier Metal Co. See Brown (1960), Jaques (1976). They are constantly under review by the social analytic research units at Brunel Institute, cf. Social Services Organisation Research Unit (1974), Appendix A.

[6]   This assumption relates to the other meaning of matrix – a medium within which an influence can develop. Latin: matrix = womb.

[7]   See Chapter 10, Note [8] for a definition of co-management.

Part II

Analysis

8 Questions about matrix management

The need for questioning

There is a danger that matrix management, like so many other useful but limited management concepts before it, might become the new panacea and flavour-of-the-month, taken up with exaggerated expectations by organisations that do not need it and are not equipped to handle it, only to be discarded with obloquy a year or two later, to join all the other good ideas that never had a fair trial. In the meantime, however, the attempt to turn an adequate functional structure into a matrix might do a lot more damage than a misdirected MbO programme.

What increases the danger is that matrix can be all things to all men. 'A very efficient way of getting things done in a hurry'[1] to the pragmatist, a sophisticated approach to complexity and environmental change to the systems enthusiast, a democratic and self-actualising way of life to the reformer, and a means of maximising use of resources to a cost-conscious board. There is of course a grain of truth in each of these expectations, or there can be, according to how a matrix structure is implemented. But it is unlikely that all of them can be met, at the same time, and to the full extent of the original hopes.

A reading of the case studies in Part I should help to guard against excessive and generalised enthusiasm. A number of these (Gunz and Pearson, Sheane, Dixon) show that managers and professionals working in and out of matrix organisations have substantial reservations about them, while Hey emphasis that much of what passes for matrix structure is not an alternative to the usual hierarchy but a special form of it, subject to the same disciplines and dilemmas.[2]

The case studies also point to the fact that matrix organisations are more appropriate to some situations than to others. Both in ICI and in the laboratories studied by Gunz and Pearson their application is partial and selective, and even in Scicon a very successful structure for consultancy projects proved unsuited to the project work of a computer bureau. At the same time it becomes very clear from the wide variety of different organisational models described in Part I that matrix management covers a range of options rather than a single specific way of structuring responsibilities. These options will be more fully explored in Chapter 10; for the moment let us note the point and return to the more general issue of what this range of options, and the approach embodied in them, might realistically try to achieve. The first questions the sceptical manager will want to ask are 'how effective is matrix management?' and 'what is it good for?' The aim of this chapter is to clarify the implications of these questions by trying to answer them.

How effective is matrix management?

Our case studies raise this question but do not answer it. Only in the consultancy case is there any clear evidence of an improvement in effectiveness. The R and D and manufacturing examples leave the matter in doubt. Were improvements due to other parallel changes? What would have happened without a matrix? The advertising study, on the other hand, suggests that the real influence on effectiveness may come, not from the structure but from the way it operates in practice. And in the public service setting, where effectiveness is notoriously difficult to measure, attention shifts to the situations which call dual structures into being.

Evidence of the effectiveness of matrix structures is on the whole rather hard to find. The literature contains a number of individual success stories, like the one documented by McCowen, which report either successful achievement of the initial objectives or quantified measures of improvement, or both.[3] These case studies can be very impressive, but it is impossible to draw any general conclusions from them. What they do not tell us is whether matrix organisations are more effective, in some definable way, than other comparable units.

I only know of two studies, both American, which provide any evidence on this. One is the Sloan School Project on the management of science and technology (Marquis, 1969) which compared the performance of thirty-eight firms working on R and D contracts for the US Government. This showed that the existence of project teams (regardless of whether they formed part of a matrix) increased the likelihood of meeting cost and time targets, but that matrix structures in which 'most of the personnel were directly responsible to the project manager for work assignments but remained physically located with their functional manager' were associated with superior technical performance in comparison with purely project and purely functional organisations.

The other piece of evidence comes from a Harvard Business School study of market-oriented programme management (Corey and Star, 1971) which, as well as a series of detailed case studies, included a survey of some 500 among the 1,000 largest US manufacturing companies in 1968. About 70 per cent of these firms had some form of product programme management, and these firms overall were more successful than the remainder in developing and introducing new products. The responsibilities of the product programme managers ranged from only being concerned with marketing and product planning to a wide-ranging co-ordinating responsibility taking in not only promotion, pricing and advertising, but also field sales, manufacturing, research and design. It was found that those businesses whose programme managers had considerable authority and responsibilities cutting across several areas (and who were thus closest to a matrix structure) were the most successful in developing and introducing new products.

Obviously, the evidence on the general effectiveness of matrix management is rather slender. But questions about general effectiveness are in any case not very meaningful. It seems to me that it makes more sense to look specifically at the things that matrix management can or cannot achieve. This means asking the second question, 'what is it good for?'

What is matrix management good for?

The question can be rephrased a little more explicitly: what are the reasons for adopting a matrix management structure in preference to any other? What can such an organisation achieve better than other alternatives? Now these are not questions, like the previous ones, that can be answered by looking for evidence in the literature or sending out questionnaires. They have to be thought through logically, deductively, to arrive at what is essentially a theoretical answer. There is nothing wrong with that – most organisational changes are based on theoretical answers to questions just like these. The idea that theory is the opposite of practice is only put forward by people who do not like the theory that is being put to them, and is one of the most threadbare excuses for avoiding an unwelcome issue. A 'practical' decision that is not guided by some form of theory, implicit or explicit, is almost inconceivable.

What is organisation for?

To make anything of these questions about matrix management, however, we have to answer some even more basic questions which are implied by them: What is any organisation structure good for? Why should one ever choose one structure in preference to another? What indeed is organisation for? Most managers, I find, have a mental block about these questions which it takes some time to remove. The mental block consists in trying to find the answer in the criteria of organisational effectiveness which they use to measure ultimate performance, and on which often their personal future depends. So their first response is to say that a good structure is one that maximises profit or return on investment, or which minimises the cost of a given standard of service, or whatever the objectives of the enterprise may be.

This type of answer is very unhelpful, I am afraid, because it over simplifies the relationship between an organisation's structure and its objectives. The first thing we need to do is to distinguish between the objectives of the *organisation* and the objectives of *organising* (i.e. organising in this way rather than in that). An analogy will help to make the point. A man is running for a bus. What is his objective? Survival? Or getting to the office in time? Excluding the possibility of his being pursued by a killer (when his objective might indeed be survival), the fact that getting to the office in time will contribute in a very small way to keeping his job, that his job is one of a number of ways of earning money and that earning money is one of a number of ways of keeping alive, is surely not sufficient reason for saying that survival was the objective of his sprint?

Compare this with another question. What is a sales director's objective in reorganising his department from a regional to a product basis? To increase sales and thereby profits? Or is it to turn his salesmen into specialists with a better understanding of their product line and manufacturing processes and how they relate to the changing needs of their customers, thus helping to improve the service they are able to give, thus helping to increase the firm's market-share, etc.? The failure to distinguish between the aims of a particular component of management structure and the ultimate objectives of the enterprise seems to me one of the biggest obstacles to using organisation theory as it could be used, to improve organisation.

Effects of structure as intervening variables

Organisation structures are arrangements for getting work done by groups of people. There are always different possibilities to choose from. Structuring the work in one way may lower unit costs. Yes, but if we do it that way we might run into morale problems leading to conflict, or we might find it very expensive to make a change in the process at a later date, if it should become necessary. So we have to decide what matters most to this organisation, at this point in time – unit cost, industrial relations or flexibility for the future? These are what one might call *intervening variables* of organisational effectiveness.

The term is borrowed from experimental science which distinguishes between independent, dependent and intervening variables.[4] If the depth of the pin-prick is the independent variable, and the loudness of the scream the dependent, there is a whole system of intervening variables, such as the degree of stimulation of the nerve endings at the point of impact, the chemical and electrical changes in nerve fibres and brain cells, the changes in the organism's state of arousal and in the condition of the vocal chords.

To ask sensible questions about the effectiveness and appropriateness of matrix or any other form of organisation, we need to restrain ourselves from jumping directly from independent to dependent variables, from structure to performance, and focus instead on the intervening variables. We need to transfer attention from the effectiveness of the organisation, to the effectiveness of organising in one way rather than another. To avoid confusion, let us call it *structural effectiveness*.

Criteria of structural effectiveness

Examples of such criteria can be found in all of our case studies. In ICI the desire to set up self-contained units comes up against the imperative to realise economies of scale, and to increase the effectiveness with which existing resources are used. One criterion that is being applied here (amongst others) is that of straight *efficiency* or level of resource utilisation. The restructuring of Scicon had as one of its major objectives the establishment of better *control* over resource utilisation and of separating *accountability* for this from the parallel accountability for achieving project objectives. One of the main issues which arises in the research laboratories in Chapter 2, as well as in the multidisciplinary teams set up in the National Health Service and the lateral groupings in Social Services Departments is that of *co-ordinating* the separate contributions of diverse specialists to the achievement of common goals. A criterion which recurs in a number of the cases is that of *adaptation* to a changing environment – the need for ICI to respond to market failures and commercial upsets, for Scicon to exploit new technologies and explore new types of business, for health service organisation to respond to increasing social complexity, for R and D laboratories to channel and disseminate new information from the environment. Frankel on the other hand sees the performance of advertising account groups as closely tied to their *social effectiveness* which is enhanced, as it is in the R and D and consultancy examples, by the increased opportunities for personal development which the structure affords to individuals.

In analysing organisations and considering proposals for change I have found it helpful to use the six criteria mentioned in these examples as a way of distinguishing between the different requirements which organisation structures have to satisfy.[5]

1 Efficiency This is a narrower concept than overall effectiveness. It refers specifically to the ratio of outputs to inputs which is the economist's definition of efficiency. Increased efficiency comes from maximising the use of available resources, increasing outputs without increasing cost, or providing a level of production or service at minimum cost.

2 Control This is the steering function. Is the organisation so structured as to be able to decide where it is going and take action to get there? The control criterion includes three parts: the ability to set objectives, the ability to monitor their achievement and the ability to take corrective action where necessary.

3 Accountability This may be seen as a part of the system for exercising control, but there is a separate point. Hierarchies function by assigning responsibility for tasks to individuals. The ability to hold people accountable for defined task areas and the achievement of specified goals is not just a way of keeping them in line, but constitutes a powerful means of motivating them to exercise discretion constructively and creatively.[6]

4 Co-ordination Because all organisations get work done by some form of division of labour, they have to have means of integrating the efforts of groups and individuals towards composite goals. It is necessary to avoid situations in which different parts of the organisation are pulling in different directions, sub-optimising or working at cross-purposes to each other.

5 Adaptation The organisation's environment does not stand still. An effective management structure has to have the capability to anticipate and respond adaptively to new and changing demands, from its clients, its providers of finance, its employees and the labour market they come from, the community and society at large. To respond quickly enough to

114

new and unexpected requirements it has to be able to recognise them, to solve problems and to innovate.

6 *Social effectiveness* As well as being a machine for performing work, an organisation is a social system. Its structure has to be viable socially as well as technically. This means that it must be able to satisfy its members' needs sufficiently to enlist their commitment to the organisation, and it must structure roles and relationships so as to facilitate co-operation and minimise harmful conflict between members.[7]

Using the criteria

The structural effectiveness of an organisation is defined for our purposes as its success in satisfying these six sets of requirements. They constitute intervening variables between its management organisation, and its performance measured against long term goals. They are not the only intervening variables that contribute to overall performance, but they are ones that are specifically influenced by the structural choices made by management. These criteria are useful in a number of ways which are relevant to the questions asked in this book.

1 They provide an approach to analysing existing organisations from the point of view of their effectiveness.
2 They provide a framework for making organisational choices which are subject to conflicting objectives.
3 They can help to indicate appropriate strategies for implementing organisational changes.
4 They constitute a convenient way of integrating the contributions from a wide range of organisation theory and research, to bring it to bear on our as yet unanswered question 'what is matrix management good for?'

Let us look more closely at these four uses of the criteria.

Analysing organisations When looking at the adequacy of an existing structure, the criteria provide a convenient check-list. They can be a useful antidote to the widespread approach of starting with the solution before analysing the problem. Matrix organisations have been introduced in situations where this essential first step had never been taken. Consequently there was no clear view of what they were intended to achieve, and as one wag put it 'If you don't know where you're going, any path will take you there'.

Choosing between conflicting objectives Although every organisation has to meet all the criteria of structural effectiveness to some extent, the means used to achieve any one of these objectives are likely to conflict with some of the others. An example of such a conflict was given earlier in the chapter. We can rephrase the choice between lowered unit cost, maintaining good industrial relations and flexibility for future change as being between efficiency, social effectiveness and adaptation. Our sales director reorganising his department is putting his emphasis on adaptation (to changing customer needs) and accountability (for specific product lines). In the process he may lose in efficiency (longer journeys, fewer calls) and in co-ordination (several salesmen calling on the same customer). Being able to express the conflicts in this form does not of course solve them, but it makes for more conscious and systematic decision making, having considered the full implications.

As so many matrix structures turn out to be a compromise between conflicting organisational objectives, this way of thinking about the decisions to be made can be very helpful in clarifying the issues.

Implementation strategies There is a simple point here: if a structure change has certain objectives (say, to improve spontaneous co-ordination through lateral interaction), then the method of implementing it should be one which makes the outcome more, rather than less, likely (joint involvement of the people concerned, rather than a unilaterally imposed solution which creates suspicion and insecurity).[8] This point will be taken up in Chapter 12.

Integrating organisation theory If questions about the reasons for adopting matrix management are theoretical questions, as was said earlier, there ought to be some theory for us to draw on for the answers. Matrix management, however, has hardly been researched, nor has it received much theoretical attention, except within certain specific theoretical perspectives.[9]

In my own view, organisation theory, past and present, and the research which it has inspired, are full of useful concepts and insights relevant to matrix management. All that is required is a common framework for integrating all these contributions. The structural effectiveness model provides such a framework, and in the next chapter it will be used as the basis for piecing together the theoretical answers to the question 'what is matrix management good for'?

Summary

Matrix management is in danger of becoming too fashionable. To guard against indiscriminate misapplication some searching questions need to be asked about it. The most obvious question – is it effective? – turns out to be almost impossible to answer, except very partially and inconclusively, but it leads on to the more pertinent question, 'what is matrix management good for?' To answer this however we need to look at the even more fundamental question of what kind of effects organisational structures can hope to achieve.

The answer which is proposed is that management structures are designed to achieve outcomes which are intervening variables between the structures themselves and the organisation's performance in relation to its goals. These intervening variables are proposed as criteria of structural effectiveness under the headings of: efficiency; control; accountability; co-ordination; adaptation; social effectiveness. These criteria can be used in analysing organisations, in choosing between conflicting objectives and in planning implementation strategies. For us they can provide a framework for the integration of the available theoretical answers to the question of the utility of matrix management, which is the task of the next chapter.

Notes

[1]   Dustin of Prudential Insurance, quoted by Perham (1970).

[2]   Jaques similarly insists that the construction and definition of lateral relationships, which includes 'dual work-group membership' and 'co-ordinated collateral teams', is an essential aspect of 'bureaucracy' (in the sociological sense, i.e. management hierarchy). The contrary impression is due to a widespread view of the bureaucratic organisation as 'a one-way downwards managerial autocracy with no upward or lateral connections' which he describes as 'fantasy' (Jaques, 1976, p. 273).

[3]   Examples are The Northern Electric Company (Lorsch and Lawrence, 1972); Ludwig, 1970; Goggin, 1974. I only know of one systematic evaluation of an organisational change involving aspects of matrix management (Dalton, Barnes and Zaleznik, 1968). The researchers were unable to obtain any valid measures of performance and could only measure changes in perceptions and attitudes, which showed partial, but not complete, success in achieving those objectives which had to do with personal involvement and participation of staff. Some groups felt more involved than before, others less.

[4]   With acknowledgements to Rensis Likert, who also uses the intervening variable concept in his theory.

[5]   This general approach was originally suggested to me by an unpublished paper by Philip Sadler.

[6]   In applying these criteria to matrix organisation I have found a constant overlap between control and accountability although in other contexts there is value in keeping them apart. In subsequent chapters I shall therefore use control and accountability as a single joint criterion.

[7]   Conflict need not always be harmful – matrix organisation can involve a 'deliberate conflict' (Cleland, 1968). This point is discussed in Chapter 11.

[8]   Cf. Argyris (1967) for an example of this.

[9]   The main examples are Galbraith (1973) who takes an information processing point of view, and Kingdon (1973) writing within a systems perspective.

9 Criteria for matrix management

Aim and approach

The aim of this chapter is to identify the requirements, in terms of what we have called structural effectiveness, that can be met by matrix management. What we are seeking is a theoretical basis for management decisions to adopt or not to adopt matrix structures. We do not have to develop a brand new theory for this purpose: the elements are already there in existing theories and research findings, though for the most part they have not been directly related to matrix management.

Because of the many and sometimes contradictory aims of different matrix structures which became evident in Part I, it is useful to classify and group them under separate headings. The criteria developed in the last chapter can provide these. They enable us to distinguish between the different requirements which matrix systems try to meet and to disentangle the various theoretical justifications which can be proposed. As a result it should be possible to reach some general conclusions about the needs which matrix management can, and cannot, fulfil and hence what are valid reasons for adopting it. In the following sections the criteria of structural effectiveness will be applied to matrix management in turn, drawing on examples from Part I and the matrix literature for concrete illustrations, and on the relevant theory and research for answers to the questions 'how can matrix meet these needs?' and 'can it do so better than other forms of organisation?'

Efficiency

Let me recall that I am here using efficiency in a rather restricted, technical sense to designate the utilisation of resources so as to maximise the ratio of outputs to inputs, and not, as is customary, as a synonym for speed, competence and deserving a large pay-rise. One argument is that all the criteria of structural effectiveness must boil down to efficiency in the end, because maximising the outputs from given inputs is what organisations are for. This is true only in principle and in the long term. In the practical short term, measurable economic efficiency can often come into conflict with other criteria. We have seen in Chapter 3 how the desire to establish clear accountability for self-contained 'businesses' within a division of ICI comes into conflict with the need to share resources and achieve economies of scale. In the same way the flexibility of the self-sufficient project group and the total accountability of the autonomous project manager has been sacrificed in many R and D establishments in favour of a mixed system which preserves the advantages of efficient resource utilisation achieved by grouping like specialists in functional departments (cf. Cleland and King, 1975, p.251).

The major contribution of management theory to the question of how to achieve high efficiency is of course the principle of *specialisation* which goes back not merely to Scientific Management or Charles Babbage but all the way to Adam Smith. One of the problems which particularly pre-occupied the 'administrative management theorists'[1] of the early twentieth century was the basis for grouping or separating activities into departments. Gulick (1937), for instance, proposed five main principles of grouping – purpose, process, clientele or material (for people-processing and material-converting organisations respectively) and geographical area. The major dilemma in most cases tended to be between purpose and process, whether to group together people and resources devoted to a common goal, or those engaged in similar activities and performing the same functions. The latter was seen as the more modern approach, which allowed full advantage to be taken of the principle of specialisation. As most matrix organisations represent a compromise between grouping by purpose (project, product) and process (function, discipline) this argument is of great significance for us.

Briefly, the efficiency criterion favours functional or process organisation because it promotes specialisation, enables work to be subdivided into elements which can be easily learned and performed more predictably, enables fluctuations in demand and in attendance to be more easily absorbed and smoothed and makes possible economies of scale through mechanisation, long production runs, ability to justify expensive equipment and hire of scarce skills. Grouping similar specialists together also fosters the development of skills, expertise and professionalism, thus upgrading the organisation's human resources. In many of the reported instances of matrix organisation the need to preserve these advantages of the functional organisation is the most powerful argument for a matrix, as against a move to a complete project or product structure. In these cases the matrix is not a means of increasing efficiency but of maintaining it at an acceptable level as well as achieving some of the other objectives we shall consider.

There are some cases, indeed, where a matrix is introduced with the express purpose of increasing efficiency. Examples of this have been developments in management consultancy (Ludwig, 1970) and higher education (Chapter 1). In these instances the original structure has been based on purpose, or area, or clientele, but with time the emerging needs for expertise in more than one place have proved too expensive for such a decentralised structure. In some cases this could lead to a radical re-organisation into a functional structure, in others the half-way house of a matrix with functional co-ordination or overlay. In these instances the matrix creates the ability to build up specialist resources, and achieve an improved, if not optimal, level of their deployment.

Many people now argue that the quest for efficiency through special-isation went too far, and that the way to improve efficiency in modern organisations is to increase individual motivation through greater partici-pation and interaction within the working group (Likert, 1961, 1967). In some approaches such as job enrichment (Herzberg, 1966) and socio-technical design of work groups (Trist et al., 1963; Emery, 1959) it is actually claimed that improvements in efficiency will result from a reduc-tion in specialisation leading to greater individual or group autonomy. Matrix organisation is seen by some writers as belonging to this group of approaches – I am sceptical about this (see section on social effectiveness and concluding section). Conversely, matrix is sometimes blamed for lowering efficiency, through increasing the time spent in unnecessary meetings and time-wasting communications. This seems to occur under rather special circumstances and is discussed in Chapter 11.

Control and accountability

Control, the ability to set objectives, monitor their achievement and take corrective action where necessary, is usually associated with systems rather than with organisation. A frequently used model is that of a nega-tive feed-back device, such as the thermostat, which monitors deviations from set limits and makes corrections, in order to bring a process back within those limits. In 'steady state' manufacturing systems, say a factory making light bulbs, such a model provides a good representation of the main control processes. The objectives set for such a system are basically in terms of resource utilisation, and control systems exist in order to maintain efficiency within the desired limits.

But systems which are involved in 'one-off' projects, in developing new products or carrying out unique tasks, have two sets of objectives and therefore require two types of controls. The resource utilisation or efficiency objectives are still important and control systems are required to look after them. But they will tend to be subordinated to the unique goals which the organisation is trying to achieve and which need in some sense to be fought for.

In the Scicon case described in Chapter 4, improved control turned out to be one of the most important outcomes of the reorganisation. But what produced this improvement was the ability to separate two distinct control systems: the one concerned with overall resource utilisation, made the responsibility of the new resource managers, and the other with the achievement of project objectives (including the control of project costs) for which the project managers were made specifically accountable. This case illustrates two important points – the need to distinguish between controls concerned with ongoing resource utilisation and those concerned with achieving specific task goals, and the close link between control and accountability.

In local government, mentioned in Chapter 1, one of the pressures towards matrix organisation seems to be the recognition that most of the control systems, as well as accountability (except where there is a scandal or disaster), relate to ongoing expenditure, i.e. resource utilisation, and hardly any to the objectives the expenditure is designed to achieve. Programme area teams are an attempt to provide a focus for the latter kind of control and accountability.

Accountability is sometimes seen as a major problem in matrix organisation – how can people operate in situations of dual accountability, i.e. accountability *to* two managers? It is important to recognise that at the managerial level another aspect of dual accountability – accountability *for* two sorts of performance – is often, on the contrary, the reason for introducing a matrix. Instead of holding the same man responsible for resource utilisation and goal achievement, often in conflict, accountability is split. In the R and D departments that use the 'leadership matrix' (Chapter 2) it is a split between the accountability for maintaining the level of professional expertise and that for completing projects within time, cost and specifications. In the ICI divisional matrix it is between production efficiency and market performance. In some social services departments it may be between providing an integrated local service and ensuring the standard of work of individual specialists.

Indeed one of the most important reasons for establishing a matrix organisation may be to create a new form of accountability that was not present before. This is illustrated by Ludwig's (1970) account of the development of a matrix structure by an aircraft manufacturer. It started because customers could not get the information they required about the state of their particular project. The company appointed first 'project expediters', then 'project co-ordinators', and finally 'programme managers'. At each stage the role acquired more responsibility and authority. The essence of this development was to establish, within the firm, proper accountability for each project, linked with the possibility of exercising cross-functional control. The functional heads were of course accountable managers, but their accountability was for resource utilisation (efficiency), side by side with a series of sub-goals, related to their portions of various projects. It was impossible to balance the two principal types of accountability, for resource utilisation and for goal achievement, within individual functions. The same problem would have existed with a complete project organisation: in either case the accountable manager has complete control over one of the elements and only partial control over the other.

What these various examples illustrate is the use of matrix organisation to resolve the basic dilemma of the administrative management theorists, between organisation by process and by purpose. While the former was seen to improve efficiency in the use of resources, the latter had the advantage of helping to concentrate efforts on a single goal and creating individual accountability. In situations where the overriding emphasis in control and accountability is on goal achievement – the success of a product, the completion of a project within given parameters – the administrative management theorists would have advocated organisation by purpose.

Current organisation research shows these dilemmas to be particularly related to increase in organisational size. The findings of the Aston group[2] and the very clear and elegant model of the stages of organisational growth distilled by Greiner (1972) from a number of case histories both tell the same story. As organisations grow larger they tend both to decentralise managerial responsibility, setting up profit-centres and product-divisions (organisation by purpose), and to increase the degree of specialisation (organisation by process) within the decentralised units as well as centrally. The need to balance achievement of new goals through autonomous, accountable management with the efficient utilisation of resources through specialisation and embracing controls is evident throughout, and leads eventually, in Greiner's scenario, to the creation of lateral groupings, teams and matrix structures.

A similar evolution is observed in a study of control systems used by different technologies (Woodward, 1970). As one moves from jobbing to mass production, controls go from 'unitary' to 'fragmented' and from personal to impersonal. In the jobbing 'unit' or 'small batch' production firm which tends to use a product form of organisation, control is exercised by the boss or product manager in person, and is a unified process, in which conflicting criteria like quality, cost and completion time are continually balanced against each other. In the functionally specialised organisation, a number of separate control systems are operated independently by different specialists, and are difficult to integrate into a total picture. In the more complex process industries, fragmented controls were found to be integrated at an impersonal level by data processing systems. Matrix management seems to provide an alternative: an organisational means of integrating resource and goal controls.

What all this amounts to is that the dual accountability of matrix management can be seen as a response to a basic split between the need to manage tasks and the need to manage resources. Normally these two responsibilities are combined in one role, but, with the increasing size and complexity of organisations, situations are arising where the resources required for a given task, or the tasks calling for a given resource, need to be managed separately. At this point a choice has to be made between concentrating on one dimension of control and accountability, leaving the other to look after itself, or using a matrix approach to focus on both.

We have begged one important question. Is the dual accountability we have talked about feasible? What about the people at the receiving end? The cases in Part I suggest that at the very least some kinds of matrix management are possible in some situations. They also show that the notion of two bosses with equal and identical authority is, to say the least, misleading. The key to the problem seems to be found in the way in which such structures are set up. Different models of matrix management involve different levels of accountability which may be complementary rather than conflicting. These are discussed in the next chapter, while the question of whether clear definition of roles in the matrix is desirable and how it can be undertaken is considered in Chapters 11 and 12.

Co-ordination

If organisations get their work done by assigning accountability for parts of that work to individuals at successive operational levels then one of the crucial tests of any organisation structure is its ability to integrate the delegated tasks and sub-tasks into the total outputs of the organisation. The criterion however is not, how *well* does the organisation integrate the work of its parts, but does it achieve the *necessary* degree of integration?

There are many different levels and forms of co-ordination. Co-ordination can mean simply ensuring consistency in the application of rules and policies in different parts of the organisation, ensuring, for instance, that a local concession to a small group of workers at one site does not create a precedent which will create problems for management at many other sites. Or it can mean ensuring consistency in professional or technical practices, seeing that financial operating statements are presented in the same form in different parts of a company, or that the customers of a bank will receive the same amount of consideration at different branches. Further levels of co-ordination include the avoidance of duplication between departments or units, and the avoidance of unnecessary or harmful internal competition, such as salesmen competing for orders from the same customer. The more complex forms of co-ordination, however, have to do with dovetailing the work of different parts of the organisation with each other, and particularly coping with those situations where a constant flow of information to and fro is required between units working on closely related parts of the same task. Hence, in applying the co-ordination criterion of structural effectiveness the important prior question is the level of co-ordination which the organisation requires.

The co-ordination objectives of matrix organisations tend to be of this more complex type. The examples of co-ordination matrix described by Gunz and Pearson in Chapter 2, the engineering matrix charged with the construction of a chemical plant (Chapter 3), and the inter-disciplinary teams in the Health Service, are all concerned with integrating the contributions of diverse specialists to complex and many-sided tasks. In the ICI example chemical, electrical and civil engineers have to work with production managers, outside contractors and finance and personnel staff, each with their own superiors and accountabilities, to erect a plant to meet certain unique performance requirements within a budget and timetable.

The administrative management theorists saw the problem as essentially one of grouping people into organisational units. If the critical forms of co-ordination could be brought under a single boss – project manager or department head – any residual problems of co-ordination between units could be taken care of by the higher levels of the management hierarchy. Increases in specialisation and complexity seem to have defeated this simple strategy in many present day situations.

Thus Lawrence and Lorsch (1967) make the point that different functions of a modern manufacturing firm may have to develop very different outlooks, processes and management structures in order to relate to different environments and this makes them more difficult to co-ordinate. In the plastics firms which they studied, for instance, the research departments operated in a very uncertain scientific environment with long time horizons, and the more effective had a rather flexible and informal structure. They found it difficult to relate to members of the production department, operating with a known technology and quick feedback on their performance within a fairly structured and formal type of organisation. The status conscious middle manager in the hierarchically structured production department does not know who to talk to in the much more casual research set-up. The researcher becomes impatient of the channels of communication he is expected to go through in his contacts with production. The general point made by Lawrence and Lorsch is that the greater the differentiation between departments the more difficult is their integration. But they point out that the need for integration would also vary in different types of organisation.

The theoretical basis for analysing this need for co-ordination is provided by Thompson (1967), writing from a systems theory perspective. He points out that co-ordination needs are based on *interdependence*, and distinguishes between three levels of this: pooled, sequential and reciprocal. Pooled interdependence is limited to the fact that both units make use of the same pool of resources, e.g. two production units serviced by the same maintenance department. Co-ordination can be achieved by a set of rules or procedures to ensure consistency and determine priorities, with possibly occasional meetings to sort out problems. Under sequential interdependence unit B cannot do its work until unit A has done its share, typically two successive stages in a production process. The preferred method of co-ordination for sequential interdependence consists of scheduling systems of various kinds. Sequential interdependencies can of course be very complex and the co-ordination systems likewise, such as some of the large networks used for scheduling complex projects.

The most difficult co-ordination problems, however, arise where interdependence is reciprocal. This is most simply defined by saying that the outputs of each unit become the inputs of the other and *vice versa*. Thus as work progresses each unit is dependent for its part of the work on information coming from the other in a continuous exchange. The methods of co-ordination used for this type of interdependence have to be ones which permit mutual adjustment of the operations of each of the units in the light of information from the other one. This kind of requirement cannot be covered by rules or by planning and control systems of a more conventional kind. If planning or scheduling systems are used they have to be highly responsive ones able to process large volumes of information and frequently to update it. The alternatives, as Galbraith (1973)

126

points out, are organisational structures which provide continuous lateral communication, either through appointed intermediaries or in the more complex and fast moving cases by continuous face-to-face contact between the people concerned.

This is one of the main reasons for adopting some form of matrix management in many of the cases which have been documented. The options discussed by Galbraith range from informal contact between managers, through co-ordinators, to the full 'overlay' matrix described in the next chapter. As a matrix structure can be demanding and difficult to operate, managers will be wise to use simpler forms of co-ordination wherever possible. The critical factors seem to be the intensiveness of the communications required and the time and resource pressures under which they have to take place. Where a lot of information has to pass back and forth between a number of participating groups under severe time constraints a matrix may turn out to be after all a simpler and more cost-effective solution than large information systems and liaison staffs.

Adaptation

The accelerating rapidity of change in our environment has become an article of faith with writers on management (Bennis, 1966; Argyris, 1972). If the environment is changing as fast as they say then organisations have got to acquire the capability of coping with this change; this is what is meant by the criterion of adaptation. Another way of putting the same thing is to say that environmental uncertainty is increasing and organisations need to be structured in ways which enable them to absorb and deal with this uncertainty. Uncertainty is defined by Galbraith (1973) as 'the difference between the amount of information required to perform the task and the amount of information already possessed by the organisation'. Adaptation can therefore be seen as a capacity for processing information.

This comes out clearly in Gunz and Pearson's survey of R and D establishments which suggests that a key task of project co-ordinators is to channel the flow of information from the environment and ensure that it is acted on. An ingenious mechanism for adaptation is incorporated in the matrix structure of Scicon. While the project managers' responsibility for repeat business ensures that information about current client needs will be evaluated and used, the resource managers have been made responsible for developing and bringing in new work, and this constitutes one of the principal means at their disposal for maintaining and improving staff utilisation, the major criterion against which their performance is judged. The structure thus facilitates the integration of two streams of information – about customer needs and about new technical developments and their possible applications. In a similar way, the need to respond to client requirements underlies the account management function in the advertising agencies described by Frankel.

Dixon describes the interdisciplinary structures of the National Health Service as a response to growing social complexity, as well as to the increasing complexity of the health care process itself, while Sheane sees the development of business area and product group organisations overlaid on a functional structure as due to the need to respond to market changes side-by-side with an expanded capital construction programme.

In these examples, adaptation is seen to involve anticipation of, and rapid response to, changes in the organisation's environment, as well as the capacity for problem-solving and innovation. While it may appear that this can be achieved by making designated individuals – the resource managers in Scicon, the account managers in the agency – responsible for monitoring and responding to change, in practice the response is almost always more broadly based, and involves expertise and creative energy from different parts of the organisation. To mobilise such a response it is necessary to bring together information from various sources, and to relate the various parts by bringing to bear on them a range of diverse and specialised insights. That these are essentially self-regulating processes which cannot be legislated for or programmed in advance by set procedures is shown by Frankel's finding that the quality of an account group's performance depended not on formal co-ordination mechanisms but on the unprogrammed interactions within the team.

These examples show matrix organisations providing a means of improved adaptation, but they do not tell us under which circumstances they are superior to other alternatives. After all there are well-known and tried methods of adaptation which are organisationally much simpler. The basic approach to the problem has, of course, always been through the traditional management functions of forecasting and planning, which may be supplemented by instituting a more continuous monitoring of key environments, such as markets, technologies, and economic conditions, so as to allow more frequent updating of forecasts and plans. Galbraith (1973) and other writers have suggested that the volume of information to be processed is getting too large for simple forecasting and planning methods to keep up with. The more complex methods that would be required may be too costly, or unavailable, so that organisational alternatives need to be considered.

The simplest of these, as already mentioned, is of course decentralisation of decision making, so that those who are in touch with what is happening in the field are also the ones to take the necessary action. Greiner (1972) for instance sees this as one of the major reasons why decentralisation at a certain stage in an organisation's development will lead to accelerated growth. 'Decentralised managers with greater authority and incentive are able to penetrate larger markets, respond faster to customers and develop new products.'

Decentralisation of management is a way of re-establishing the *status quo*, in which managers can manage autonomously, simply by reducing the scale and complexity of the problem. An alternative prescription, for situations where the information to be evaluated and responded to is too specialised and diverse for the individual accountable manager, is Burns and Stalker's 'organic organisation'. The key factors in this seem to be high commitment to organisational rather than sub-unit objectives, a great deal of flexibility about the nature of individual contributions to the task in hand, authority based on knowledge rather than status, but above all constant and intensive personal interaction to ensure that necessary information gets to the people who need it and can respond to it. In the small electronics firms described by Burns and Stalker these requirements were apparently met fairly easily simply by letting a highly motivated group of people get on with the job with a minimum of structure or directions. Some large organisations have drawn the moral of this observation by setting up small autonomous 'venture groups' or 'blue sky units' to develop new products or markets without the organisational weight of the large parent body. There is evidence from a number of researchers that in uncertain and changing situations loosely structured, more flexible organisations perform better than those which emphasise rules, standardised procedures, strict separation of tasks and precise definition of roles, while conversely the latter organisations do better in stable and predictable situations.[3]

In systems theory terms (and it is necessary at least to mention systems theory, as matrix organisation is sometimes described as a direct expression or application of a systems approach – e.g. Gray, 1974) management has to work with two contrasting models, a 'closed system' model for efficient control of the 'core technology' and an 'open system' model for adaptation to the environment.[4] In the former the system boundary is strengthened and defended through various methods of buffering outside perturbations (inventories, forecasts, contracts, restrictive practices), while the latter requires a permeable boundary through which goods, services and information flow to and from the environment in a constant exchange.

In stable situations performance is maximised by strong boundary control, because with low uncertainty the management hierarchy can act as a filter and translate information from the environment into plans and programmes for efficient operation. But an uncertain, rapidly changing environment generates too much information and clogs up the filters. At this point it becomes necessary to resort to open system logic, by making the boundaries more permeable so that the operating systems which produce goods and services can interact directly with the environment. This is what organic organisation seems to be about.

The alternative possibilities of forecasting and planning, decentralisation and the small organic organisation indicate the circumstances under which matrix organisation might represent a more appropriate way of meeting the adaptation criterion. Where a large organisation is operating in a diversified environment subject to rapid change, and is not able to decentralise its operations into self-contained units (whether because of scale economies or internal interdependence), or meet its adaptation needs by hiving off autonomous units concerned with innovation, a matrix structure is often seen as securing some of the advantages of both decentralisation and organic organisation. The creation of accountable project or product managers provides a means of delegating decision making to those responsible for the critical interface with the environment, such as the area managers in social services or the business managers in ICI.

In systems terms the matrix seems to provide a way of reconciling open and closed system logic. Functional or process departments remain the guardians of core technologies, with control over their own boundaries, which is, however, mitigated by the superimposed horizontal structures which ensure that these boundaries are kept sufficiently permeable to permit the necessary information flow to and from the environment and the process of ongoing adaptation to environmental demands.

The matrix creates crossing communication channels (in place of the totally unrestricted ones of the organic organisation), and, like organic organisation, it emphasises horizontal communications, commitment to total rather than partial goals and 'authority' based on ability to contribute to problems rather than rank. In so far as the project or programme manager can ensure free and flexible interaction of members of different specialist groups within the team setting, the conditions existing in an organic organisation should be reproduced. In fact the matrix team is seen as an encapsulated organic organisation.

One problem about this view is of course that the project team in a matrix is not and can never be encapsulated and the whole point of setting it up is that it should not be. This in turn means that it can only operate organically if the organisational cross currents to which it is exposed will allow it to. Some reports from the field have shown that often matrix organisations do not work like this but get bogged down with formalities of various kinds because, as Argyris (1967) has underlined, they are superimposed on a culture and a set of behaviour patterns based on the clearly marked boundaries and restricted information channels of a more rigid organisational tradition.

Social effectiveness

The criterion of social effectiveness, the idea that the organisation structure should provide for a viable social system which satisfies the needs of its members, sounds a very simple and straightforward one. In reality it is far from simple and can be used to support a number of very divergent and even contradictory measures. What makes the idea of social effectiveness sound so simple on the surface is its link with the equally simple notion of the 'socio-technical system' proposed by researchers at the Tavistock Institute (Emery, 1959). This suggests choosing between the different possible arrangements of the technical system for getting work done in such a way that the resulting social relationships both support the technical objectives and contribute to the satisfaction of the social and psychological needs of members. Within the framework of socio-technical analysis, usually applied to groups of manual workers, the analyst tends to apply a widely agreed list of specific socio-psychological needs to be satisfied by work. But when the notion of social effectiveness is related to the wider problem of organisation design the decision maker, or the theorist, can take his pick among a range of different 'needs' leading to different conclusions. The variety of the social needs which matrix organisations sometimes fulfil, and sometimes fail to satisfy, comes out in the case studies. All of these raise issues of social effectiveness, but not always the same ones.

At the individual level Gunz and Pearson, as well as McCowen, report that matrix organisation increases opportunities for personal growth and development, both by creating additional managerial jobs and by increasing the visibility of the high performer.

The difficulty of belonging to two working groups, one of them temporary, is raised in contrasting ways in two of Gunz and Pearson's examples. In one case staff with strong loyalties to their functional departments apparently found the setting up of project groups with even as long life spans as two to three years too temporary and hence unsettling and this was given as a reason for abandoning the matrix structure. In another laboratory by contrast the home-bases of the technologists were described as 'vestigial' and clearly did not inspire any feelings of group identity, resulting in a wish to perpetuate the project groups by not concluding projects. In Scicon on the other hand the creation of ongoing resource groups as well as project groupings has been a stabilising influence and has apparently contributed to reduced turnover. In general it seems that creating specialist groupings superimposed on a project structure provides staff with a source of group identity,[5] while the reverse process, of adding project, product or business groups to an existing functional structure often increases insecurity.

Another source of discomfort comes from uncertainty in role relationships, a problem which seems to crop up in matrix organisations in our R and D and manufacturing examples as much as in the interdisciplinary health teams. An approach to resolving some of the uncertainty by patient elucidation of these relationships is illustrated in Hey's analysis of dual influence and accountability in social services departments.

According to Gunz and Pearson the project manager in certain types of R and D matrix has an important leadership role, in creating a cohesive group of committed and motivated members. But many people now believe that leadership is not enough, and that commitment to the organisation and its objectives requires the ability to participate in decisions which affect one's work and hence feeling personally responsible for their outcomes. This is one of the powerful ideological reasons for advocating matrix organisation, as Dixon, as well as Frankel, points out. The interfunctional working parties set up in social services departments are directly inspired by such a desire to foster participation in policy making, and in the advertising agencies studied by Frankel, high performance was associated, amongst other factors, with the degree of individual participation in the work of the group, reflected in the measure of 'member centrality'. The remaining studies, however, do not provide evidence of higher individual participation being fostered by matrix structures, and the evidence that this is a general result of introducing matrix management is by no means clear-cut, though the literature contains some examples (e.g. Northern Electric in Lorsch and Lawrence, 1972; TRW in Rush, 1969).

The widely held view that matrix organisations enhance individual satisfaction, motivation and commitment to organisational goals is based on a picture of the matrix team as an autonomous, flexible and participative group of equals. There is quite a lot of evidence that membership of a group with agreed objectives and free interaction between members helps to give a sense of identity and belonging and satisfies important psychological needs.[6] It is also a source of motivation, but whether it motivates members to work for or against organisational goals seems to depend on how they define their common interests, how the group is integrated into the organisation and the extent to which members share in decisions about their work. The work of the Tavistock Institute using the concept of the 'socio-technical system' has already been mentioned. Typically, the practical application of this concept by Tavistock and other action researchers has resulted in the formation of 'semi-autonomous work groups', in which the members have a joint responsibility for a complete and meaningful task and are themselves in control of the allocation of work within the group. This type of structure has been found to increase group identity and sense of belonging as well as motivation and commitment.

Matrix organisations are often spoken of in the same breath as autonomous work groups, both by enthusiasts and sceptics, and supposed to have some of the same effects. Within interfunctional groups, such as project teams, advertising account groups or business area teams, relations are supposed to be non-hierarchical and based on knowledge rather than formal authority. Members are motivated to work together towards a common objective and participate in the decisions that govern their work. In some situations this may indeed be the case, but there seem to be a number of obstacles to realising this ideal.

Firstly, as the chapters by Gunz and Pearson and by Sheane make clear, the dual group membership characteristic of the matrix is disliked by many of the members, precisely because it is seen to run counter to the simple cohesion and identity of the single product or discipline-based group. A key feature of the semi-autonomous work group is that its members are buffered against external interference in the detail of their work by having a clear boundary between themselves and the rest of the organisation. This is a luxury which the product group or project team in a matrix cannot afford if its purpose is to permit the interaction of the resource groups, knowledge bases and information streams of the functional structure. Kingdon has noted that good lateral communications in the project team tend to go with poor vertical communications between team members and their functional superiors, and *vice versa*. This highlights the difficulties of a 'team' which has to operate as an open system. Each member is both in the group and outside it, and if it is not to seal itself off, and cease to fulfil its integrating role, the team can become a stage on which the conflicts inherent in the surrounding situation are played out. Yet often this is not an aberration, but a necessary condition of the team's existence. The Sloane researchers (Marquis, 1969) noted that the highest standards of technical excellence were achieved by those matrix organisations in which most of the members of the project team spent part of their time in their own functional departments.

Another problem is mentioned by Dixon at the end of Chapter 6 – it seems not to have been noted elsewhere. The chances are that an interdisciplinary group will contain members of differing capacity and level of work, differences which may negate the theoretical assumption of equal participation in decision making within the team. Some will be more equal than others – the example of doctors and nurses merely dramatises what is probably a much more general problem which our current ideologies tend to obscure.

Probably the most one can say in confirmation of the idealised picture of matrix as a more participative and 'democratic' organisational form is that some of the situations which are pushing organisations towards matrix structures may also be creating pressures for greater participation and delegation. It seems unlikely that a change in structure on its own will bring such changes in management style into being, but the view of many who have experience of matrix organisation is that the one will not work without the other.

Thus, the verdict on the social effectiveness of matrix structures is very mixed:

> they can provide social support and group identity where it is lacking, but they can also threaten and undermine the same qualities in an existing set-up;
> they create uncertainty about individual roles and relationships, but provide opportunities for personal growth and development lacking in simpler organisations;
> they facilitate collaboration between departments and individuals, but are also subject to a great deal of conflict;
> they can increase participation in decisions, personal motivation and commitment to the organisation but are not bound to do so.

Matrix organisations seem to be more difficult to live with and work in than stable functional hierarchies. Indeed, it appears that they require a good deal of social sophistication from their members to function at all. But set against this is the widespread perception that, in situations of high reciprocal interdependence on tasks with exacting targets, the matrix creates a social system which matches the task requirements better than other forms of organisation. On this reading if matrix is difficult to operate in, that is because of the inherent difficulty of the situations which lead to its adoption. It is a form of organisation which provides more scope than most, both for conflict and stress, and for personal commitment and creative collaboration. Much may depend on the existing organisational culture and on the ways in which the structure is introduced and operated.

Criteria for matrix management

Having considered the relevance of matrix management to each of our criteria of structural effectiveness we are not left with a simple conclusion, or the ability to set out one definable set of circumstances to which matrix provides a best fit. Instead we find that there are a number of organisational needs to which matrix management may provide an answer, though in most cases it will not be the only answer, and we therefore need a rationale for guiding what will never be an automatic decision. In any case such a decision will only be the choice of a general approach; we still have to decide on the precise form of organisation, out of a number of alternatives, in which this approach is to be embodied.

Our survey showed that most matrix structures are introduced with more than one objective in mind. From a list of eight possible aims, two-thirds of the respondents picked out at least three, while a third checked five or more. Priority was given to objectives associated with efficiency, control and accountability, and co-ordination.[7] As I see it there are, talking in rather general terms, three good reasons and one somewhat doubtful one for wishing to adopt matrix management. Each of these reasons may be of overwhelming, strong, moderate or negligible importance for the organisation in question, which suggests a cumulative choice model, rather like adding weights to one pan of a scale until it tips. On the other side of the scale are the costs, economic, social and psychological, of a change to matrix management. And, finally, there is one additional weight, representing the rather vague concept of 'organisational culture', which may go on either side of the scale, either facilitating or inhibiting the change. The possible reasons for wishing to adopt a matrix approach emerge from the foregoing analysis. So also, at least implicitly, do the factors which will determine the importance or 'weight' of each reason for the organisation in question.

Combining efficiency with goal accountability

One reason is the wish to combine the efficiency, economies of scale and level of expertise of a functional structure with the task-centred accountability and control of project or product organisation. The starting point is most commonly a functional organisation into which it is desired to inject project or product accountability, but the converse can also occur where the aim is to raise the functional efficiency, resource utilisation and expertise of a project, product or area organisation. The importance of this particular aim will derive from the extent to which the tasks and goals for which accountability is to be established are central to the organisation's role or critical for its survival, together with the pressure on the organisation's resources, the presence or absence of 'slack', and the extent to which a lowering of efficiency is acceptable.

Reciprocal interdependence

Another motive may be the existence of high reciprocal interdependence (cf. p. 126) between organisationally separate and differentiated units or even individuals engaged in common tasks. The factors which would give high weight to this reason would be the pressures, financial, technical and particularly time, to which the work is subject – the greater such pressures, the stronger the case for matrix management.

Adaptive potential

The need to increase the organisation's adaptive potential in respect of innovation and response to uncertainty and rapid change is the third indicator for a possible matrix structure. The weight of this reason seems to be related to the number and distinctness of the separate environments, or aspects of the environment, whose change and uncertainty presents problems for the organisation, as this determines the number and diversity of information flows and internal contributions to be integrated.

A more 'democratic' organisation

A further reason which has sometimes helped to motivate the adoption of matrix management is the desire to replace a hierarchical management system, perceived as authoritarian, by a less hierarchical one perceived as more democratic. The weight given to this particular reason will depend very much on the power and prestige of the main advocate of the change and that of his most committed converts.

Some doubts about the realism of these expectations have already been expressed, and the ways in which matrix structures may redistribute power within organisations will be discussed in greater detail in Chapter 11. Some redistribution of power is indeed likely to result, as with most reorganisations. For this to be in the direction of a more democratic system requires special provisions, such as those made by the Northern Electric Company (Lorsch and Lawrence, 1972), otherwise some people may well participate in more decisions than before, and others in fewer. Whether democracy is an appropriate model for a work organisation (cf. Jacques, 1976, pp. 200-2), and whether this type of exercise does not simply represent an imposed substitution of one value system for another are both questions one could argue about, and which help to explain why I described this fourth reason as dubious. Nevertheless it may, in conjunction with some of the other reasons, help to tip the scales in favour of matrix management in some actual situations.

Effects of culture

The effect of an organisation's 'culture', that is to say its prevailing values, beliefs and methods of operating, is either to facilitate a move towards matrix management because the latter implies behaviours and relationships which are already normal and preferred in the organisation, or, alternatively, to inhibit it because behaviour and attitudes would have to change more or less drastically as a result. In the latter case the cumulative weight of the reasons in favour will have to be greater than where the culture favours the change.

The analysis of organisational culture is not yet a very well developed field. Handy (1976) has proposed a useful classification, distinguishing between task cultures, power cultures and role cultures. The task culture emphasises commitment to the job in hand and tends to prefer informal collaborative relationships. As a result it fosters attitudes and behaviour which are very favourable to matrix management. By contrast, the role culture, which is very prevalent in large and highly structured organisations, emphasises rules, procedures and clear role-definitions, and the existence of such a culture might well weigh on the opposing side of the scale. The power culture tends to centralise direction and decision making in the hands of a few influential individuals or a single top manager. Such a culture might weigh either way according to whether the matrix structure was itself 'plugged into' the main power source (e.g. by the chief executive also being its originator) or whether it was being proposed by people in a position of little power.

Summary

Assessing matrix management in the light of the existing body of theory and research on the management of organisations, it has been possible to identify some of the most important ways in which matrix structures could be expected to affect criteria of structural effectiveness.

Efficiency A matrix may help to maintain existing levels of efficiency when project or product groups are introduced into a functional organisation. It can be used to increase efficiency where important resources are dispersed between sub-units. But there is also evidence that matrix management sometimes lowers efficiency when introduced into rigid and formalised organisations.

Control and accountability A matrix is often a recognition of the need to have two forms of accountability and control, related respectively to the efficient use of resources and the achievement of specific task goals. Thus the dual accountability of the matrix organisation is not one of its unfortunate by-products, but a major reason for its existence.

Co-ordination Matrix structures are among the most far-reaching and expensive methods of inter-unit and interpersonal co-ordination, which tend only to be justified in situations of high reciprocal interdependence requiring constant mutual adjustment under strong time and technical pressures.

Adaptation In situations of rapid environmental change and uncertainty, matrix structures seem to recreate within the large, highly-structured organisation some of the means of adaptation which have been identified in simpler contexts. Their main strength in this respect lies in their ability to facilitate the exchange of information and responses laterally between departments and different specialists.

Social effectiveness Matrix structures can be difficult social systems for individuals to work in. While the resemblance of some matrix teams to semi-autonomous work groups may give them the social advantages of the latter in the form of increased participation and commitment, they can also generate stress and confusion owing to conflict and ambiguity in individual roles and multiple group membership.

The analysis suggests that there is no single set of circumstances to which matrix management provides a best fit, but that its choice is likely to be based on a cumulation of requirements and the weights given to them. The existing organisational culture may also play an important part in the decision whether to embark on a matrix structure. Such a decision, once taken, is however only the first step – it still remains to identify the precise form which the structure is to take. To examine the options available for this is the task of the next chapter.

Notes

[1] The 'administrative management theorists', sometimes also known as the 'classical' management theorists, are usually held to include Taylor, Fayol, Gulick, Urwick, Brech and some others. They are for the most part managers generalising from their own experience in order to formulate a set of prescriptive 'principles' of good management. While they have been criticised for making assertions which are not backed up by systematic research evidence, they raised many of the main issues of concern to organisation theory, and provided us with a set of concepts for discussing organisational problems. Summaries of their writings can be found in Massie (1965), Kast and Rosenzweig (1970), Pugh et al. (1971).

[2] The 'Aston group' is a loose designation for a group of researchers at the University of Aston and later the London Business School and a number of other centres, led initially by Derek Pugh, who have over the last fifteen years carried out a very exhaustive series of large scale comparative studies aimed at relating measurable aspects of organisation structures to their context and ultimately to organisational performance and behaviour. A large number of references could be quoted, but these have recently been collected together into Pugh and Hickson (1976), Pugh and Hinings (1976), and Pugh and Payne (1977).

[3] This is probably the strongest and most consistent finding of the now substantial number of studies carried out within the general perspective of contingency theory, i.e. the theory that different types of organisation structure are appropriate to different contexts and situations, e.g. Burns and Stalker (1961), Woodward (1965), Lawrence and Lorsch (1967), Harvey (1968), Aiken and Hage (1970), Child (1974, 1975).

[4] This particular dilemma was identified and fully explored by Thompson (1967).

[5] This point is more fully developed by Miller and Rice (1967) who distinguish between 'task groups' and 'sentient groups'. Where tasks are of limited duration followed by regrouping, the task group does not provide long term social support and identity, i.e. it does not constitute a 'sentient group' and there is a need for more permanent professional or functional groups to provide members with an organisational 'home'.

[6] This evidence comes mainly from the work of the 'human relations' school, from the Hawthorne investigation onwards (Roethlisberger and Dickson, 1939), the work of the Survey Research Centre at the University of Michigan (Likert, 1961, 1967), and work at the Tavistock Institute of Human Relations (Trist and Bamforth, 1951; Rice, 1958). Some indirect evidence is also provided by the many investigations of the effects of organisational size on indicators of morale and satisfaction, which is reviewed by Porter, Lawler and Hackman (1975, pp. 249-52).

[7] Respondents were asked to select those aims which led to the setting up of the matrix and rank them in order of priority. They had the option of describing the aims in their own words or selecting from the following list – most of the former could be classified with items in the list and are included in the following figures, which give the proportion of about sixty respondents selecting each aim, and its average ranking.

139

	per cent	mean rank
Improve utilisation of certain resources	46	1.8
Maintain utilisation of resources while achieving other objectives	12	2.1
Concentrate resources on high priority tasks	32	2.9
Increase control over/accountability for projects/ products/businesses/markets	43	2.4
Improve co-ordination and integration of different groups or departments	43	2.7
Improve ability to respond to change/innovate/ problem-solve/cope with uncertainty	49	3.1
Increase level of specialist expertise	34	2.6
Create a more participative/democratic/motivating/ satisfying organisation	29	3.4

10   A choice of models

Matrix organisations have developed from different antecedents and in different contexts. In the last chapter we saw that they can also serve a variety of purposes. It is hardly surprising therefore that in practice they take many different forms. In surveying the matrix structures found in research and development laboratories, Gunz and Pearson were able to distinguish between two principal types, with a number of individual variants. A similar variety of structures is observed by Sheane within a large manufacturing group, and by Hey in local authority social services departments. The aims of this chapter are therefore firstly, to review the main models of matrix organisation and some of their many variants, and secondly, to apply the criteria already established to the more specific choices between types of matrix.

Clarifying role relationships

Although the different models of matrix will be described in rather broad terms it is useful to be more precise about the nature of certain of the key role relationships. A very powerful approach to achieving such precision has been developed by Elliott Jaques under the title Social Analysis.[1] This involves an analyst working collaboratively with people concerned in organisational problems to clarify and formulate important aspects of the existing and 'requisite' situation. One of the very useful outcomes of social analytic research has been the precise formulation of major role relationships which are commonly found in organisations. In my opinion these provide an invaluable means of elucidating some major options in complex organisations and are particularly useful in defining the different possibilities available in a matrix structure. A number of these relationships are referred to in this chapter and in the chapters by Dixon and by Hey, both of whom have worked as social analysts.[2]

Three main approaches

There seem to be three principal approaches to matrix management, each of which produces a typical organisation structure which has a number of variants. As there are no generally accepted names for these approaches I have had to select my own. They are:

the co-ordination model
the overlay
the secondment model [3]

In the *co-ordination* model, staff remain organisationally and managerially members of their original departments, usually functional, but procedural arrangements are instituted to ensure cross-departmental collaboration and interaction towards the achievement of extra-departmental objectives. In the *overlay*, staff explicitly and officially become members of two organisational groupings, each with its own manager, while the *secondment* model describes a situation in which individuals move from functional departments into project groups and back again, but belong to one or the other at any one time. The term matrix organisation is sometimes restricted only to the overlay model but the broader term matrix *management* tends to include all three, and the distinctions between them are not nearly as radical as appears at first sight.

The co-ordination matrix

This seems at present to be the most widespread form of matrix management. About half the reported matrix organisations in our survey were of this kind. There are plenty of examples of different types of co-ordination matrix among the case studies in Part I: R and D departments, advertising agencies, aspects of health and social service organisation, and probably some parts of ICI.

The philosophy underlying the co-ordination approach is that it is better not to change the organisation any more than can be helped, and that ambiguities and conflicts in reporting relationships are to be avoided. This is accompanied by a belief in the power of social skills and willing collaboration to take the place of explicit structural change.[4] One of the problems of the co-ordination matrix may well be, as the case study by Argyris (1967) illustrates, that the organisations most likely to go in for this option are also the ones in which inter-departmental collaboration is the most difficult to achieve.

In a typical project co-ordination matrix the basic organisation is functional but there are a number of project co-ordinators (they may be called project managers) each of whom is responsible for a particular project and its successful completion within given cost and time targets and quality standards. For each project there is likely to be a project team

142

which meets periodically, whose members are in frequent touch with each other, and which is chaired, rather than managed, by the project co-ordinator. The team is made up either of all the people working on the project in the different functions, or, in the case of large projects, of their section leaders or immediate managers. In some cases the team consists of designated 'representatives' of the participating departments who themselves merely play a monitoring or co-ordinating role in relation to project work in their own department – a particularly weak arrangement in which the project co-ordinator can easily be sabotaged by the appointment of low-calibre 'representatives'. A co-ordination matrix for product, rather than project, co-ordination is found in some manufacturing firms.

Co-ordinating roles are very widespread. Jaques summarises the definition as follows:

> . . . the person in a co-ordinating role is accountable in relation to the task concerned for: negotiating co-ordinated work programmes; arranging the allocation of existing resources or seeking additional resources where necessary; monitoring actual progress; helping to overcome problems which may be encountered; reporting on progress to those who established the co-ordinating role.
>
> In carrying out these activities the co-ordinator has authority to make firm proposals for action, to arrange meetings, to obtain first-hand knowledge of progress, etc., and to decide what shall be done in situations of uncertainty, but he has no authority in case of sustained disagreements to issue overriding instructions. Those co-ordinated have always the right of direct access to the higher authorities who are setting or sanctioning the tasks to be co-ordinated. (Jaques, 1976, pp. 268-9)

This does not seem to go as far as the full project responsibility (for time, cost and quality) which is frequently assigned to project 'managers' who are really co-ordinators. There are three possible views of such a responsibility–authority gap:

1   The project manager/co-ordinator is not really held accountable for achieving project objectives, providing he alerts higher management of any insurmountable (for him) obstacles to their achievement early enough for something to be done.

2   The project manager *is* held accountable, but disposes of real power in excess of his manifest authority.[5]

3   The project manager is accountable for results, but has no extra power or authority: he has an impossible job.

Half-hearted attempts to operate co-ordination matrices on the basis of the third alternative are probably among the experiences that have given matrix management a bad name.

Variants of the co-ordination matrix

The collateral team

In this model there are no designated co-ordinators, but interdepartmental project groups, working parties, teams or task forces are jointly charged with co-ordinating their own, or their departments', work on each project or area of operation. Accountability for portions of the work will tend to rest with individual department heads and their staff, the members of the team being responsible only for ensuring that they get, and give, the necessary information required by each department to do its part of the job. Meetings of the team may be chaired by one of its members, and as is evident from Dixon's description of collateral teams in the Health Service, this may raise expectations of 'leadership', unjustified in this context.

Where the people, the project and the environment are right, and the 'team' is given the necessary leeway by all the department heads concerned, such a group can sometimes become much more than a series of information-exchanging meetings and acquire a real dynamic of its own, to become in fact something like the semi-autonomous work groups of socio-technical theory. But there is no reason to expect this to happen in the majority of cases – in most organisations there are probably more factors working against such an outcome than in its favour.

Co-ordination by stages

The co-ordinator role can pass from person to person as the project moves through a series of stages. This makes sense where different stages of the project are dominated by different functions or specialities. Really this means that the project is split into a number of sub-projects, each with its own co-ordinator. Bergen (1975) in his description of a project information system gives a glimpse of this approach in operation.

Project expediter

In the advertising account groups described by Frankel the role of the co-ordinator falls to the traffic manager who acts as project expediter. In other contexts this is sometimes known as the 'staff' model, a somewhat weaker form of project co-ordination. In this form the project is monitored on behalf of the responsible senior manager by what is often a staff assistant whose responsibility is to keep himself informed of all aspects of the project, to point out to the individual managers involved any failures to meet project objectives and standards, and to report on progress to his superior, who will take any decisions that are required. Owing to his information role a project expediter is likely to represent a 'node' in the communication network for the project, and as such will fulfil an important co-ordinating function in fact, if not in theory.

Customer-contractor model

Kingdon (1973, pp. 138–43) describes how under strong pressure from functional managers the system for controlling certain projects was transformed into the 'work-package' relationship which, in extreme cases, deprived project managers of direct access to the functional staff working on their projects. This is an example of the customer–contractor model, a variant of the co-ordination matrix which really retains very little of the character of matrix organisation.[6] In this situation the project manager, probably supported by his own nucleus of full-time staff, negotiates for work to be done with the individual functional managers, whose relationship to him is a 'service-giving' one. The project manager is in control of the budget for the project and can therefore decide the allocation of project finance to the individual departments, stipulating in exchange completion dates and standards to be met. In this model, although functional managers retain full managerial control over their contributions to the project, the project manager is theoretically in a position to take full responsibility for the outcome.

In practice there is, however, an important difference between most such arrangements and a true customer–contractor relationship, because the project manager is not usually a free agent able to take his custom elsewhere if he is not satisfied. A customer–contractor model therefore risks leaving him with the worst of both worlds – the 'arms-length' relationship deprives him of the real co-ordinating power of direct involvement with the participants, while his theoretical authority to negotiate and insist on fulfilment of the 'contract' leaves him in the position of responsibility when things go wrong.

Product management

Where product managers are appointed in a functionally organised manufacturing firm, and are charged with watching over the performance of given products, including product development, launching, specifications, price, quality, packaging, distribution, sales volume, market share and profitability, the true situation seems most often to be a co-ordination matrix of the first type, weakened occasionally to the 'expediter' function.[7]

Davis (1973) gives an interesting and detailed account of the introduction of a product management structure into the international division of an American company organised into four geographical areas. Four product managers (described as matrix managers) were appointed in order to co-ordinate product ranges across areas. Each product manager adopted a different strategy, one concentrating directly on the sales performance of the foreign subsidiaries in his product group and a second attempting to take over the broad marketing role of the area managers for his products, while a third focused on increasing the international penetration of new developments in the domestic product division.

The situation can be very different in the advanced technology research and development industries such as aerospace. Here most products *are* projects and many of the reasons may exist which push project management beyond the co-ordination matrix, as in the example of an aircraft manufacturer described by Ludwig (1970) which was quoted in Chapter 9.

Case co-ordination

Another form of co-ordination matrix is sometimes said to exist where the focus of the work of a number of organisationally separate specialists is a single person (or a series of such persons) such as a patient in a hospital or a client of a social service agency. In many cases one individual takes on the co-ordinating role for each client. This has been found to be a nurse in a psychiatric hospital organised as a therapeutic community (Stamp, 1974), or a field social worker in a social services setting (Social Services Organisation Research Unit, 1974). In these cases the co-ordinator's lack of managerial authority is obvious and hence expectations may be more realistic.

The overlay

This type of structure (exemplified by Gunz and Pearson in Fig. 2.3), unlike the co-ordination matrix, represents a definite and deliberate departure from traditional models of organisation. It flouts the principle of unity of command dear to administrative theorists like Fayol (1949), it challenges the view that people working in organisations are too inflexible to accept managerial authority from more than one source or to share managerial accountability for the work of joint subordinates.

The emphasis in the overlay structure is on ensuring a simultaneous two-way flow of information, between specialists within a project, and between projects within a specialism. It is also on maximising at one and the same time the accountability for goal achievement of the project manager and for resource utilisation of the functional department head. The aim is to equalise the authority and power of the project and functional managers, and while this may be, as Galbraith (1973) puts it, 'an unachievable razor's edge', and most actual examples of overlay structures seem to tilt a little either towards the co-ordination matrix or towards the secondment model, the principle of striking a balance between two equally valued sets of objectives remains an important one.[8]

In the typical project overlay, the functional organisation has superimposed on it a project structure, made up of a number of groups, each headed by a project manager and composed of members of the appropriate functional or specialist departments. The relationship between the individual member of a project group, his project manager and his functional head has been identified by social analytic research as co-management.[9] The specialist manager (here the functional head) agrees the member's allocation to the project with the operational manager in charge, remains responsible for appraising his technical competence and for helping him with technical problems. The operational manager is responsible for allocating work and for both monitoring and appraising its performance. The relationship depends on the continued agreement of the two managers, or on the existence of a 'cross-over point' manager who can settle disagreements that may arise (the 'top left-hand corner' position in the matrix), but if such disagreements become numerous the overlay will become unworkable.

The exact allocation of authority and responsibility between the two managers will obviously vary from one organisation to another, and the reality of the relationships established may vary from project to project and from one individual to the next. What tends to be constant in most attempts to operate an overlay model is the effort to spell out in as much detail as possible the respective authority and responsibility of project and functional managers.

147

Variants of the overlay

Sub-project managers

This version of the overlay restricts dual reporting relationships to relatively few people and among reported cases it seems to be more common than the full overlay embracing all project staff. Three of the fullest descriptions of matrix structures in the literature are of this type, which seems to be particularly appropriate to project or product groups involving large numbers of staff.[10] Under this system each specialist department contributing to a project appoints a sub-project manager in charge of a sub-group within the department, working on that project. The project staff within the sub-group are managed only by the sub-project manager, but the latter is accountable for their work both to his departmental head and to the project manager. Dual accountability is thus confined to a single level of the hierarchy. TRW sometimes combined this system with the practice of having individual members of staff contributing to the work of more than one sub-project group.

Dual roles Gunz and Pearson (Chapter 2) mention a practice which seems to have been adopted by a number of research establishments. The activities of the establishment are organised in grid fashion, with specialisms being overlaid by projects, but the managers of the two groupings are the same individuals. Thus, adapting Fig. 2.4, the role of project manager for project I might be held by the head of department A, which also does most of the work on that particular project.

This points to the fact that the idea of the 'interdisciplinary' project is not typical of most types of project activity. In many research laboratories projects are centred on individual disciplines with only subsidiary contributions, if any, from other specialisms. This suggests a 'project within function' model (see p. 151) with subsidiary contributions on a 'service-giving' basis. Why then go to the trouble of setting up an overlay with what Wilkinson (1974) describes as 'two-headed' managers?

In the case in question the motive is the achievement of a balance, not of authority, but of emphasis on two kinds of objectives. In his functional manager role, the head of department A is concerned with the development of a long term research programme in his field of application. But the money for such a programme, under the company's existing policies, must come from projects commissioned by the manufacturing divisions. Hence in his role as project manager he is responsible for earning the money which he and his colleagues need to maintain and develop their departments as viable research resources for the company. This is not

148

simply a peculiarity imposed by one company's accounting procedures. Few companies these days are prepared to finance pure research divorced from visible application, and the organisational balance between developing a resource and achieving project objectives is one that most laboratories have to maintain in some way.

The business matrix

ICI's development of matrix structures which overlay 'businesses' or business areas on existing functional organisations is described in Chapter 3, which also indicates how an additional geographical dimension of integration has to be added at the corporate level.

Large diversified international companies, embracing multiple technologies, production facilities, product groups and markets, tend to use a variety of organisational forms, including variants of the co-ordination matrix and the overlay. Two recently reported examples are Corning Glass and Dow Corning. In the former a system of seven 'world boards' responsible for the company's major product groups as world-wide 'businesses', and chaired by business managers, has been overlaid on a profit-responsible geographically based area management structure. Thus at the highest level the chief executive of a subsidiary in one country is responsible to both his world board and his area manager (Hill, 1974). Dow Corning has gone a step further by establishing what it calls a 'multi-dimensional' organisation. The basis of this is a system of ten business boards regarded as profit centres overlaid on a structure of five main functions described as cost centres. While the business manager is only in a co-ordinating role to the functional members of his business board, the business board itself is jointly accountable for the profitability of the business to the corporate business board which includes all the function heads, thus effectively creating an overlay situation at senior management level, based not only on individual managerial relations but also on the relations between boards at different levels. The additional dimension of a geographical organisation comprising five areas has been added to this structure but the resulting managerial relationships have not been described (Dow Corning Corporation, 1973; Goggin, 1974).

Functional overlay

A business organised entirely into project or product groups, each with a full complement of the necessary specialists and resources, might at a given point in time find it possible to make savings or, more positively, to build up expertise and co-ordinate policies in major areas of specialisation by grouping specialists from the different projects into functional groups overlaid on the project structure and appointing managers, not to run projects, but to build up and improve the utilisation of the specialist resources of the company in those areas (cf. Videlo, 1976). A similar situation arises in geographically based organisations, such as consultancy firms operating from a series of area offices.[11] Consultants who are specialists in a particular field thus become members both of an area office, working for local clients, and of a specialist team providing support and development in their field and allowing their deployment outside their home territory where necessary.

Functional overlays also seem to be taking the place in some firms of what is generally known as the 'dotted-line relationship' between specialists such as personnel officers and accountants at individual sites and more senior specialist managers located centrally. Where the dotted-line relationship meant keeping in touch, giving advice and encouragement and sorting out the occasional spot of bother ('functional monitoring and co-ordinating' in the social analytic vocabulary), the use of the functional overlay means that the site personnel officer or work study manager can be both a member of local management, and part of a company-wide, centrally-managed personnel or consultancy service.[12]

The secondment model

In most cases where this model is used, the functional structure is seen as a service to what is essentially a project-centred organisation. The philosophy of the secondment matrix tends to be that the project manager must be in full control of all his more important resources, and that there should not be any ambiguity about who is in charge. But side-by-side with this conviction is the perception of a set of needs which the project structure by itself is unable to meet. There needs to be a more permanent link between project staff and the parent organisation, to avoid a mass-exodus when a project appears to be nearing its end. Individual specialists have a need to belong to a group made up of colleagues in their own discipline, not just psychologically or to maintain their expertise, but also to provide them with a career ladder. Equally the organisation may feel the need for its own pool of specialist expertise, rather than going to the open market at great expense every time a new project comes along.

In the typical secondment matrix such as the one at Scicon (Chapter 4) functional departments exist side-by-side with full-time project groups with individuals moving back and forth between them. Most of the project staff retain nominal membership of their functional department

150

and automatically return to it when no longer required on a project. At the same time a large or small proportion of work for projects is done within the functional departments on a service or customer–contractor basis.

An extreme form of this model is described by Kingdon (1973, p. 137) as the 'body shop'. With this the project managers were able to demand, and to obtain, the services of named individuals from the functional departments at very short notice, and to return them to their departments equally abruptly. The functional manager was left to carry the costs of the disruption, the project manager only being charged with the time actually spent in the project group. This extreme proved to be too disruptive, and over a period of time the functional managers, with the support of senior management, were able to redress and indeed reverse the power balance to achieve the 'work package' arrangement – a customer–contractor form of the co-ordination matrix.

The name I have adopted is based on the fact that 'secondment' is the relationship identified by social analytic research which seems to predominate in this model of the matrix. This is described as involving the transfer for a limited period of a subordinate from his original manager to a new superior who takes over almost all managerial responsibility and authority, leaving the original manager accountable for a continuing official appraisal of the subordinate's work, for his formal training and for making plans for his career development.

Side-by-side with secondment, this type of matrix tends to make extensive use of 'service-giving' where the service-giver (a functional manager or specialist) remains fully accountable for the work done, once requirements for the service to be provided have been negotiated by the project manager.

Variants of the secondment model

Project within function

In the research laboratory of an electronics company each project is firmly located in one discipline-based department which provides the main inputs to the project. The project manager reports to the department head, and inputs from other departments are provided either on a service-giving basis, or by occasional secondment of specialists from other disciplines. This is a very useful variant of the secondment model, because it is a way of achieving its objectives without any need for matrix structure at all.

Like the dual role variant of the overlay, it applies to projects where most of the work is done within a single functional department, and its use is limited by the size of the contributions required from other departments. If these are extensive and require close and ongoing integration with other parts of the work, one could imagine the project within function rapidly turning into a full secondment matrix with the project manager's accountability to the head of one of the functions involved becoming a complication rather than a simplifying factor.

Functional location

A way of attempting to combine the advantages of the overlay and the secondment model is to formulate managerial authority and accountability in terms of the latter, while leaving the members of the project group physically located in their respective functional departments (cf. Videlo, 1976). The danger with this apparently brilliant compromise is that secondment without a physical move may come to appear totally unreal, and that the physical proximity of the functional management could undermine the theoretical authority of the project manager, and reduce the real interaction between members of the project group.

Core group

In this version the bulk of project work is done in the functions, with only a small full-time team reporting directly to the project manager. Hence this model involves a combination of the secondment model with the co-ordination matrix or the overlay. In one version, this core group, which may function as a 'project office', is made up of specialists seconded from the functional departments involved in the project, who help to monitor and co-ordinate work being done in their home departments. Alternatively, full-time project staff may, in Sayles and Chandler's term (1971, p. 183), be 'colocated' in functional departments to participate in the work being done by departmental staff working on the project.

Choosing a model

It would be dishonest to pretend that there are any grounds other than theoretical and common sense ones (one hopes they are the same) for suggesting which of these models are most suited to given situations and objectives. The following remarks must therefore be regarded as speculative and hypothetical – they make sense to the author and are not contradicted by what little data are available, but they have not been put to any sort of test.

From the many relevant considerations which will vary from case to case, let me emphasise four which seem to apply fairly generally. These are:

1   the antecedent organisation;
2   the organisation's 'culture';
3   the people available;
4   the objectives to be met (using criteria of structural effectiveness).

Antecedent organisation Starting, as is most common, with a functional organisation, the easiest and most natural development would be towards a co-ordination matrix. Any attempt to jump straight to one of the other forms would require drastic changes in attitudes and behaviour, and would be perceived as very threatening by existing managers. If the point of departure is a project organisation, the natural first move would be towards a secondment matrix or a functional overlay.

Culture Using Handy's (1976) classification again, it seems that a role culture (emphasis on rules, procedures, role definitions and boundaries) would have difficulty in operating the overlay model, not because either the accountability or the authority of the managerial roles cannot be spelt out as clearly as can be desired, but because of the generally rather formal and impersonal relationships valued in such a culture, which would tend to get in the way of the highly unprogrammed forms of interaction which the overlay calls for. Handy's task culture on the other hand, with its emphasis on autonomy, flexible relationships and goals, should take very readily to this form of organisation. A power culture (emphasis on central control and personal authority) might have difficulty with both the co-ordination matrix, except perhaps in its project expediter form, and the group autonomy required by the overlay.

People In general, by introducing an additional set of managerial roles, a matrix organisation calls for general management ability which is often in short supply at lower levels. This need will be particularly strong in both the overlay and the secondment matrix. Co-ordinating roles with clearly de-limited accountability may be easier to fill.

Objectives The main objectives, in terms of our criteria of structural effectiveness, of adopting matrix management structures emerged at the end of the last chapter. While the decision to adopt a matrix is often the result of a combination of several of these aims, each of them, and the priority given to it, has a bearing on the choice of a particular model.

Combining efficiency and goal accountability

All of the models discussed achieve a compromise between efficient resource utilisation, and task-centred control and accountability for project or product goals. The choice between them depends on emphasis. The co-ordination matrix leans towards resource utilisation, by maintaining the authority of the functional heads whose responsibility it is. The secondment matrix leans in the direction of project accountability by giving full control of project staff to the project manager. The overlay, theoretically, strikes an equal balance between the two.

Within these major types, the variants we have discussed also favour one or other of these two objectives. Within the limits of the co-ordination matrix the strongest accountability is vested in the co-ordinator in frequent contact with a team, while the collateral team carries much more diffuse responsibility and the expediter and customer–contractor model occupy an intermediate position. In these cases greater task accountability, even if co-ordinative, inevitably places more restriction on the functional head's ability to optimise the deployment of his staff. In the overlay, the 'dual role' model places greater emphasis on resource utilisation, while the sub-project/product manager variant concentrates project accountability in fewer hands. Of the variants of the secondment model the project in function has the best of both worlds but is only possible under certain circumstances, while the use of functional location increases the potential for high quality inputs to the project but can weaken the project manager's effective authority. The core group version weakens the control of the project manager and enhances the functional head's ability to deploy resources efficiently.

Co-ordination

Where the emphasis is on the co-ordination of inputs subject to reciprocal interdependence as defined in Chapter 9, the most important requirement is for the existence of a team in frequent face-to-face contact, and the available evidence indicates that most organisations choose a version of the co-ordination matrix, unless other considerations carry considerable weight.

Adaptation

The overlay seems to be the model which maximises adaptive and innovative potential, by increasing the flow of information within and between both functional departments and project groups and by facilitating flex-

154

ible and unprogrammed interaction between individuals. Approximations to this situation are achieved by the collateral or co-ordinated team in the co-ordination model, and by the use of functional location of staff in the secondment model.

Social effectiveness

Although some scepticism was expressed about this objective of matrix organisation, certain social criteria can have a bearing on the choice of matrix model. Collaboration and understanding between specialist groups can be enhanced by any of the models which bring people together in a team setting, while increased participation and commitment is generated by membership of a fairly autonomous team, which is most likely in the overlay. Psychological security through membership of a permanent professional group can be provided by moving from a purely project organisation to a secondment matrix or functional overlay.

Summary

This chapter reviews the basic types of matrix organisation and some of their main variants.

The *co-ordination matrix* tries to interfere as little as possible with the existing structure, as well as avoiding ambiguities in reporting relationships. Hence the horizontal aspect of the matrix is represented by monitoring and co-ordinating rather than managerial roles. There are a number of variants, ranging from the *collateral team*, a team without a co-ordinator, to the *customer–contractor* role, a co-ordinator without a team. *Product management* and *case co-ordination* are often versions of the co-ordination matrix.

The *overlay*, which involves a radical departure from the usual principles of organisation, tries to achieve an equal balance between managerial accountability for resource utilisation and for achieving project or business objectives, and tends to involve detailed specification of responsibilities along both axes of the matrix. Variants such as the use of *sub-project managers* and the *dual role* model depend on particular circumstances, while more broadly the overlay is being applied in the cross-functional or inter-area *business matrix* and in the integration of staff services in a *functional overlay*.

The *secondment model* is the name chosen to describe situations where staff alternate between membership of a project group and a functional department. It tends to be used where project objectives are paramount and the functional structure exists as a fall-back and a service to the project organisation. The variants mentioned all involve a strengthening of the functional structure in comparison with the basic secondment model.

The choice between these models depends on many considerations, including the antecedent organisation, the organisational 'culture', the availability of people to fill the new roles, and particularly on the emphasis which is placed on the different structural objectives to be met by the matrix organisation.

Notes

[1] The method is described in Brown and Jaques (1965, pp. 29-47), and discussed in Rowbottom et al. (1973, pp. 276-300) and Rowbottom (forthcoming).

[2] A useful summary of the main role relationships identified by social analytic research appears in Jaques (1976), Chapter 17 and detailed definitions are given in the appendices to Rowbottom et al. (1973) and Social Services Organisation Research Unit (1974).

[3] By a co-ordination matrix I mean the same as Gunz and Pearson in Chapter 2. Their other main type, the leadership matrix, can usefully be split into two distinct forms – the overlay, and the secondment matrix. Both of these involve a strong project manager role, but dual group membership is simultaneous in one case, sequential in the other. Other attempts at classification include Steiner and Ryan (1968, pp. 7-11) who distinguish between three types of project management – pure, matrix and influence. Sanders (1976) makes a fourfold distinction between wholly pure, dominantly pure, dominantly influence and wholly influence projects. Cleland and King (1975, pp. 252-62) distinguish between a number of specific models rather than general types.

[4] Some other assumptions underlying the co-ordination matrix are set out by Gunz and Pearson on p. 27. My description is complementary rather than contradictory to theirs.

[5] See Chapter 12 for a discussion of the possible sources of this power.

[6] Our original concept of matrix organisation as implying both dual leadership and dual group membership excludes this variant altogether. It is mentioned here as a clear alternative to the co-ordination matrix, not very far removed from it, in the way in which the project co-ordinator's responsibility and authority are defined.

[7] For a discussion of various forms of product management, see Fulmer (1965), reprinted in Cleland and King (1969) and Hopkins (1974).

[8] One company which designs and builds flight simulators operates an overlay structure, but it is found in practice that the early design stages of the work are carried out in the functional departments with the project manager acting as co-ordinator, while in the construction stage he becomes the manager of a team working together on the actual equipment (Paskins, 1977).

[9] A widely acceptable formulation of the respective responsibilities of the two co-managers with respect to a joint subordinate 'X' has been produced by recent social analytic studies at Brunel. To my knowledge this is the first general formulation of what many people regard as the 'pure' matrix situation, and it is worth quoting in full.

156

The functional co-manager is accountable in respect of X:

for helping to select him according to professional criteria, and for inducting him in matters relating to the field concerned;

for helping him to deal with technical problems in the field concerned;

for co-ordinating his work with that of other similar participants in the field;

for keeping himself informed about X's work, for discussing possible improvements in standards with him and for reporting to the operational co-manager any sustained or significant deficiencies or lapses from established policy in X's work;

for appraising his technical competence;

for providing for his technical training.

The operational co-manager is accountable in respect of X:

for helping to select him and for inducting him in operational matters;

for assigning work to him and for allocating resources;

for appraising his general performance and ability.

Both co-managers have right of veto on appointment, right to provide official appraisals, and right to decide if X is unsuitable for performing any of the work for which they are accountable.

The functional co-manager can give instructions provided that:

they are given within policies established by the 'crossover' manager, binding on both co-managers; and

they do not conflict with policies or operating instructions issued by the operational co-manager.

Since the functional co-manager is accountable for X's functional competence, he must have the authority to monitor the operational co-manager with respect to policy in the functional area, to ensure that X's competence is being utilised in a professionally appropriate way.

[10] Rush (1969) describes matrix management at TRW Systems, one of the large aerospace development contractors to NASA, as based on sub-project managers in each department. Galbraith (1971) gives a detailed account of the 'Standard Products Co' 's evolution from a functional to a matrix structure, involving sub-product managers accountable to both a functional head and a product manager. Videlo (1976) paints a more complex picture with functional section heads under department heads and project co-ordinators under high-level project managers, and an attempt being made, not always successfully, to group under a section head staff working on only one project, or at least on projects controlled by the same project manager, to improve the chances of agreement on staff deployment.

[11] E.g. 'Hickman Associates' in Lorsch and Lawrence (1972), and PA as described in Ludwig (1970).

[12] An example of 'functional monitoring and co-ordinating' is given by Hey in Chapter 7. In respect of the financial controller of a subsidiary, Davis (1973) makes a very convincing case that, even in the most traditional hierarchy, the controller has always been in a position of effective dual accountability to the local manager and his own functional superior with the dotted line to either one or the other being turned into a solid line whether by common location or by the prescribed reporting system.

Part III

Operation

11 Problems of matrix management

The matrix situation

Before looking at the problems that matrix organisations suffer from it is worth reminding ourselves that such structures are attempts to solve problems, not to create them. It is therefore at least possible that some of the difficulties which are found to be associated with matrix management and which are liable to get blamed on it are in fact inherent in the situations that make the matrix approach relevant in the first place.

Take, for instance, the problem of dual accountability. As has been shown, this arises from a need to reconcile conflicting requirements – for instance the requirement to meet certain project deadlines with the requirement to utilise skilled manpower as efficiently as possible. One approach is to place both responsibilities on the same shoulders, resulting not so much in schizophrenia as in concentration on one objective at the expense of the other, according to the way in which the control and reward systems happen to be weighted. The other approach is to make different people accountable for the two requirements in the hope that this will make it easier, not more difficult, to reconcile the conflicting objectives. Not only may the conflict which the structure appears to have generated have been inherent in the situation, but if it seems to be giving serious trouble (and not merely upsetting someone's sense of managerial propriety), the chances are that higher management has got it wrong, because the problem has been accentuated rather than solved.

In a previous paper (Knight, 1976) I noted a strange phenomenon about the reported benefits and problems of matrix organisations. It appeared that in many cases, rather than being complementary to each other, these were quite clearly contradictory. For example, where one firm reported that matrix permits a more efficient use of resources, another complained that it is unnecessarily costly. One rather obvious explanation did not strike me at the time. Assuming both firms were faced with the need to manage demanding projects within a functional structure, one has been able to use the matrix as a way of solving the problem of efficient resource deployment in spite of the project pressures, while the other has not, and is blaming matrix management for its failure to combine efficient resource management with project control.

But to say that all the problems associated with matrix management are problems of the pre-existing situation which matrix has failed to solve would be an overstatement. In practice they are a mixture, resulting partly from external circumstances and partly from the matrix approach itself. In so far as that approach consists of attempting to achieve diverging objectives by making different people accountable for them, it encounters at the outset the problem of achieving a *balance* between the respective power and authority of the managers charged with the two sets of responsibilities, a problem of organisation design and role creation. At the interpersonal level this may lead to problems concerning the *relationships* between people acting out what Cleland calls 'the deliberate conflict'. These may be accompanied by problems of *individual* adaptation to an ambiguous and potentially stressful situation. At the same time matrix brings with it additional *administrative* problems to which the previous system may not be geared.

In this chapter problems of matrix management will be considered under these four headings.

Balance problems

Sheane, describing business area matrix structures in ICI, sees the balancing of business and functional influences as a major difficulty to be resolved. The problem of striking the correct balance between two types of objective and, more tangibly, between the influence and authority of the individuals responsible for their achievement is probably the basic problem that has to be solved in setting up and operating a matrix organisation. Often the problem is implicit rather than manifest and can easily be overlooked, yet many of the more obvious difficulties such as dysfunctional conflict, individual stress, rigidity, inefficiency and failure to meet objectives may spring from it.

Striking the 'correct' balance here does not necessarily mean achieving a power equilibrium between the managers on the two axes. Writers who insist on the importance of equalising the authority and influence of the two types of manager tend to be those who define matrix in the restricted sense of what we have described as the overlay. This is a type of structure which is most often set up when task and resource objectives are given equal weight, or when co-ordination and adaptation criteria require that individuals should maintain an equal degree of involvement in their project and their speciality.

But if one accepts the broader concept of matrix which we have been using, then there are many situations in which it is wholly appropriate

162

and requisite for one type of manager to command greater authority or control of resources. Finding the correct balance in such cases consists of relating the effective influence vested in the respective roles to the aims to be achieved. Thus the multi-project setting of TRW which provides the context for Kingdon's book may well have been one where the overriding authority of project managers to acquire and release staff belonging to functional departments corresponded to a very high priority being placed on meeting project targets. If, as Kingdon indicates later, the situation changed towards greater emphasis on limiting costs, this might explain the fact that the functional heads were able to reverse the power balance by moving from the 'job shop' to the 'work package' system. The latter increased the authority of functional managers by giving them far more control over the work to be carried out on the project managers' behalf, and improved their ability to achieve economic deployment of their staff.

The balance problem has been represented by Wearne (1970) in a diagram shown in Fig. 11.1. The angle between the peak of the matrix and the horizontal represents the degree of decision making authority of the project managers in relation to functional heads. In the various versions of the co-ordination matrix which have been described the angle would seem to approach zero, but in practice some decisions are almost certain to be taken by the project co-ordinator, just as in the secondment matrix, with an apparent ninety-degree angle (complete project manager authority), certain technical and staffing decisions are likely to remain the responsibility of the department heads.

Where a matrix structure evolves in the course of time there is a chance for the organisation to feel its way towards an appropriate balance between process and purpose. A bigger problem arises where a matrix is deliberately introduced to supersede a one-way structure. Organisations where this has happened often report considerable opposition from the previous line managers, who see their authority threatened or eroded by the new structure. What sometimes fails to be recognised in these situations is that the creation of new horizontal leadership roles, whether of project managers or functional co-ordinators, must also change the content of the existing managers' jobs, and that the two sets of roles need to be defined, or redefined, in relation to each other if an appropriate balance is to be achieved. Some approaches to this are discussed in the next chapter.

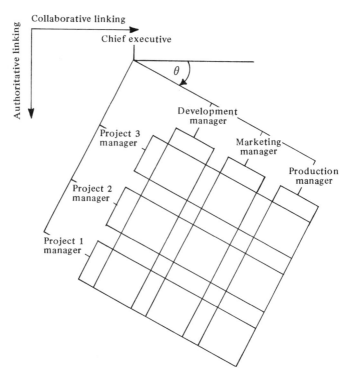

Fig. 11.1   Balance of authority in matrix organisation (Adapted from
S. H. Wearne, *Management Decision*, 1970)

Relationships in the matrix: The problem of conflict

The paradox of some matrix organisations is illustrated by an example
described by Argyris (1967). Inter-departmental product planning and
review teams were set up in one company to improve communications
and collaboration, but resulted in increased suspicion, defensive behaviour
and conflict. Argyris felt that the new organisation was justified but the
managers' style of behaviour, based on the previous structure, had not
changed to match. This he ascribed to the authoritarian and non-partici-
pative way in which the system had been implemented.

Most types of matrix organisation depend for their success, like the
advertising account groups in Chapter 5, on the willingness of their
members to engage in a great deal of unprogrammed informal interaction
outside the normal reporting channels. In some organisations this comes
naturally (in such organisations it tends to be said that 'we have always

operated as a matrix without using the name'), in others it is very difficult to achieve. It is in the latter kind that matrix is not only difficult to introduce in the existing culture, but often makes matters worse by creating new conflicts.

The case studies in Part I, of R and D departments, manufacturing divisions and health service teams, support the view of many managers in our survey who report that conflict is one of the greatest problems of matrix organisation. The conflicts most commonly referred to are those among project (or product) managers competing with each other for resources in the functional structure, and those between the managers on the two axes of the matrix (e.g. project managers and department heads) whose responsibilities overlap. In contrast, however, Cleland (1968) describes matrix management as a system involving 'purposeful or deliberate conflict' in which the integration of the contrasting functional and project interests is achieved by negotiation.

This points to the fact that organisation structures, by defining relationships, also define areas of conflict. Apart from so-called personality clashes, which are probably rarer than many managers suppose, most of the conflicts found in organisations are at the same time based on genuine conflicts of interests and are due to the way the organisation is set up. 'Divide and rule' is not only a strategy for governing colonial empires but an accurate and concise description of all organisational management. Conflicts between departments, such as production and sales, are commonplace in the traditional functional organisation. The way in which responsibilities are allocated and performance assessed defines interest-groups which are both interdependent and in competition for power and resources (Seglow, 1974). Matrix organisation tries to deal with a more complex set of requirements by superimposing a second set of responsibilities. To the structural conflict between departments it adds two further types of conflict, that among projects, products or other horizontal groupings, and that between the horizontal and vertical components. In themselves these forms of conflict are necessary accompaniments of a more complex structure, providing motivation, energy and a system for dealing with diverging requirements; it is only their dysfunctional aspects which present a problem.

Conflict becomes dysfunctional when it delays decision making, reduces significantly the amount of energy available for relevant work, blocks important communication channels and subjects individuals to unacceptable stress.[1] The extent to which these things will happen could depend on how well a system of matrix management can be integrated with the existing organisational culture.

Thus in organisations where personal advancement depends on internal patronage, making the right impression on the right people (the 'power culture' – Handy, 1976), a good deal of energy could be diverted into in-fighting if it is felt that important dispensers of patronage are not greatly committed to, or interested in, the matrix system as such, but only in certain limited outputs from it. This is the type of situation in which the open 'confronting' style of conflict resolution advocated by Sheane and certain American writers (e.g. Lawrence and Lorsch, 1967) is least likely to be effective.

Blocks to the free information flow which matrix structures depend on can be created by managers who feel that the matrix constitutes a threat to their autonomy and power. They see the boundaries of their sphere of control overlaid by the new horizontal groupings and react defensively by withholding information and emphasising formal communication channels, as happened in the cases observed by Argyris.

Individual reactions to conflict vary a great deal of course. Some find exhilarating a level of conflict which others would regard as intolerable. The special problem of the conflicts created by a matrix structure is that very often the persons expected to resolve the conflict between two managers may be those subordinates who have been made accountable in two directions. This can work if the subordinates are allowed considerable discretion by their managers, but in an authoritarian environment it can become intolerably stressful. Under these circumstances there will be strong pressure for detailed codification of relationships, ultimately the best cure for all forms of role conflict. In practice however there will be a limit beyond which the exercise will become self-defeating. Human organisations are systems for coping with reality by taking decisions in the face of the unexpected. Any attempt to pre-programme these to such an extent as to eliminate all potential conflict will remove the essential human discretion on which the system's ability to react to changed circumstances ultimately depends.

Individual problems

One of the most common problems to be reported by managers of matrix organisations is that of insecurity, stress or dissatisfaction in individual employees owing to the uncertainty of their reporting relationships.[2]

Matrix organisations undoubtedly place bigger demands on individuals, the foremost among them being the greater complexity of their reporting relationships. These may indeed be made even more uncertain by leaving

166

roles deliberately ambiguous to increase the flexibility of the organisation.[3] Approaches to role definition in matrix organisations are discussed in the next chapter, but at this point it is worth noting that ambiguous roles have been shown independently to be a major source of personal stress. A wide-ranging investigation in the United States (Kahn et al., 1964) showed that both the experience of stress and the presence of stress-related symptoms or disease were higher among people experiencing 'role conflict' or 'role ambiguity' in their work. Role conflict is the state of being exposed to conflicting expectations from the various people one works for or with so that one cannot satisfy them all, while role ambiguity is a condition in which other people seem to have no clear expectations about one's role, a state of affairs which, surprisingly, had an even higher association with stress than did conflict. Both conflict and ambiguity are frequent in matrix roles which involve the individual being a member of more than one working group and having to meet the demands of more than one superior. The problem seems to be worse when no attempt is made to define roles and responsibilities in specific terms.[4]

It may be that this problem is more serious than many managers realise. To be successful in managerial roles carrying considerable discretion requires a high level of the quality known to psychologists as 'tolerance of ambiguity'. People who have this tolerance tend to be rather impatient of attempts to spell out the details of authority and responsibility; they prefer to 'play it by ear', to construct their role as they go along. But among their subordinates there may be those who do not share their own ability to cope with uncertainty, who may be made profoundly uncomfortable, in some cases physically ill, by not knowing where they stand. The problem can be intensified if, as is often the case, matrix operation leads to an increase in individual work loads.

Another relevant consideration is the distribution of individual differences in tolerance of ambiguity. Research on adolescents has shown that 'convergent thinkers' are more likely to choose careers in science than 'divergent' ones, and the convergent/divergent dimension seems to be closely related to tolerance of ambiguity (Getzels and Jackson, 1962, p. 126; Hudson, 1963). This might mean that among scientists there are more people who find it difficult to operate with an undefined and ambiguous set of role relationships, which would be important for matrix organisations in research and development establishments. But the fact that individuals do differ a great deal in their ability to cope with ill-defined roles must be stressed. The more 'entrepreneurial' character often seeks out undefined and ambiguous situations which he can mould to his requirements, and for such people the matrix may provide great opportunities for rapid personal development and advancement.

Administrative problems and costs

On paper, matrix organisations are flexible, organic and informal, as well as fast-moving and efficient. In practice it seems that some are, and some are not. Inevitably they require people to spend more time in communicating with each other than the normal single channel hierarchy. Communicating takes time from other activities and therefore has a cost. The theory of matrix is that most of these communications will be oral and informal, that the time they take will not be excessive and will be time well spent for achieving work objectives. In some cases, such as the example described by Argyris, the opposite seems to take place. The organisation gives rise to an increase in formal communications, through meetings and memos, many of which are concerned with trivialities or with individual managers' attempts to cover themselves against blame, by putting everything in writing. A more open system of communications can be perceived as a threat and stifled in red tape. Even without such pathological reactions, a matrix system undoubtedly requires more information to go to more people. Some increase in administrative costs is likely, as well as an increase in the time spent at meetings. This sometimes leads to a feeling that the organisation is too elaborate, and to the question, 'do we really need all this organisation?' (Perham, 1970).

The other area in which administrative costs are likely to be incurred is that of support systems. Information and control systems need to be modified to fit the structure and, as Hopwood shows in Chapter 14, to be widened in their scope. Job evaluation, grading and appraisal systems may require adaptation. These changes are considered further in the following chapters, but their cost must not be overlooked.

Finally, the creation of additional co-ordinating or managerial roles also costs money. In situations which are not sufficiently complex to require them such 'integrators' can indeed get in the way of effective operation (Lawrence and Lorsch, 1967, give examples of this). As, I think, Argyris once remarked, a co-ordinator can be a man with a vested interest in keeping people apart.

Summary

Not all the problems associated with matrix organisations are necessarily caused by them; some may be inherent in the situations which lead to the adoption of dual structures.

Striking the correct balance between the authority, influence and objectives represented by the axes of the matrix is a fundamental problem, but the correct balance is not necessarily an equal one. Conflict, which is built into matrix, and indeed into any form of organisation, does not necessarily present a problem; purposeful and dysfunctional conflict need to be distinguished. Conflict and ambiguity in individual roles and reporting relationships are major sources of dissatisfaction and stress in matrix organisations, and can be under-rated by those whose tolerance of ambiguity is high. The administrative problems and costs of matrix management are related to a larger volume of communications, to the need for additional support systems and managerial roles, and, in some cases, to the defensive reactions of managers who feel threatened by the change.

Notes

[1] 'Significantly', 'relevant', 'important', 'unacceptable' – the words used are deliberately indeterminate to underline the need for managerial judgement in the actual situation. Determinants of conflict in matrix and ways of dealing with it are discussed by Wilemon (1973) and Thamhain and Wilemon (1975).

[2] Out of about sixty respondents so far to the second stage of our survey, over 40 per cent reported this problem at a level regarded as either serious or worrying – a higher proportion than for any of the other problems listed.

[3] Goodman (1967) claims that deliberate ambiguity in the definition of project managers' authority is a way of adapting to the problems of dual authority structures. See Knight (1977) for a discussion of this view.

[4] Preliminary results of the second stage of our survey show that the problem of insecurity, stress or dissatisfaction, related to uncertain reporting relationships, is serious enough to cause concern in about half the organisations which do not specify role relationships in writing, as against only a quarter of those that do.

12 Managing matrix: some approaches

Looking for answers

It is easier to state problems than to provide answers. Not enough is known about the operation of matrix organisations, or indeed of complex organisations generally, for concrete workable solutions to their problems to be proposed with any show of confidence. But at least I think we know roughly where to look for the answers to some of the major ones. Let me start by sketching out a number of approaches which look as if they ought to help with the problems examined in the last chapter.

To establish the right balance between the managers on the vertical and horizontal axes of the matrix it is necessary first to pick the most appropriate organisational model for the situation and objectives, as discussed in Chapter 10. On this basis the respective roles and responsibilities can be defined; whether this should be done with great precision or in a way which leaves a lot of room to manoeuvre needs to be considered. By itself this will not resolve the problem; if the required balance has been determined there remains the need to provide the managers in the system with the necessary power to meet the responsibilities placed on them.

Next there is the problem of dysfunctional conflict. To come to grips with this, the issue of personal and departmental power has to be faced again, but this time in the context of internal 'politics'. The method adopted to implement the matrix structure can help either to exacerbate or to neutralise such power struggles. The question of how, and how much, to define roles arises again. The ability to work together in cross-functional teams and resolve conflicts of objectives can be improved by training and organisation development approaches, such as group relations seminars and team building. The methods and criteria for judging individual and managerial performance and the information systems which give expression to these criteria are also important.

The individual stresses and uncertainties of working in a matrix are again related to role definition and performance criteria, as well as to reward systems. The need to develop an understanding of the more complex pressures and relationships of a matrix organisation, and for skill in coping with them, is a challenge to the trainer. In addition the knowledge that individuals differ in their ability to tolerate uncertainty must be taken into account in the selection of staff for both managerial and operating roles.

Some of the administrative problems and costs of matrix management may be unavoidable, and they underline the importance of questions such as: Do we need a dual structure at all? How little co-ordination can we get away with? The ability to resolve the earlier problems of balance, conflict and individual adaptation will undoubtedly affect the ease and reliability with which people can work together in the matrix without elaborate administrative fail-safe devices, and the method of implementation may set the tone for future operation.

Clearly a number of themes recur in this catalogue of approaches to matrix problems – the issue of ambiguity and of defining roles, the question of power and its distribution, the methods used in implementing a matrix, the training and team building activities which can assist it, and the support systems that have to be provided including not only management systems of information and control but also systems for selecting, appraising and rewarding staff. Although each of these topics could be dealt with at great length, our present knowledge would hardly justify it, and we shall therefore survey them rather briefly in this chapter.

Ambiguity

Whether one feels that 'ambiguity' in relationships is to be avoided or that 'flexibility' of interaction is to be encouraged is basic to one's approach to managing matrix. Unfortunately the two aims, though each laudable in itself, can be seen as contradictory. It is partly a matter of culture: task cultures prize flexibility, role cultures dislike ambiguity, while power cultures may not care very much either way. It is partly a matter of objectives; adaptation or co-ordination of intensive reciprocal interdependence call for a great deal of unprogrammed and hence flexible interaction, while efficiency and accountability tend to require clear definitions. It also depends on the people involved; some prefer uncertainty, others cannot live with it. The fact that there is often some correlation between culture, people and objectives can be a help in practice, but the top managers, whose views on the matter will be decisive, must not overlook the fact that their own tasks and temperament may be very different, i.e. more 'flexible', than those found at the operating levels.

The view taken on this issue is most forcibly expressed in the approach to the setting up of horizontal roles, such as project or product managers. There seem to be three main approaches:

defining both responsibility and authority

defining responsibility but not authority
leaving both undefined

Though there are plenty of strong views among managers and theoreticians about the rightness or wrongness of each of these approaches, I do not think any of them is necessarily correct under all circumstances. The first approach depends on a patient process of analysis and clarification of role relationships, while the second relies on the assumption that, in the absence of formal authority, the project or product manager can be given real power and resources to enable him to meet the responsibilities placed on him. The third approach will be viable only where there is a great deal of commitment to, and unanimity about, organisational objectives and a sensitive understanding of how they can be achieved.

The ways in which these three requirements can be met are discussed respectively in the following sections on defining role relationships, on power, and on organisational development and training.

Defining role relationships

The traditional organisation chart, with its set of positions joined by lines, carries clear implications about authority and accountability relationships which are well understood by the people involved. They realise, of course, that life is not as simple as the chart makes out, but that does not invalidate it; the chart is the skeleton of accountability which is given flesh by a host of subsidiary relationships. The job descriptions which often supplement such a chart tend to be much more peripheral to the work being done; they are needed for administrative purposes like selection, training or job evaluation, but it is unusual for them to be closely studied by the job holders themselves, except when they are feeling defensive or bloody minded.

But, as Hey shows in her second figure, the simple organisation charts of yesterday have acquired a distressing tendency to become covered with dotted lines, whose implications are much less clear. A further step away from clarity is taken when the 'tree' is represented by a grid, with lines crossing each other at right angles. Should the meaning of these dotted and crossing lines be spelled out? That is really what the argument is about – not whether everyone should have a detailed job description. The alternative is to leave the people concerned to sort it out for themselves – let them decide what the dotted lines mean, not in so many words, but in the ways in which they choose – or do not choose – to relate to each other.

That is what Burns and Stalker meant by an organic organisation – can it work in a matrix?

We have noted the view that to reduce ambiguity one has to sacrifice flexibility, but it is just possible that this view could be based, at least partially, on a misconception. This is the idea that by defining relationships between roles one is constraining initiative, reducing the scope for creative interaction. Certainly, definitions can be used in this way, but they need not be. The more constructive view is to treat them as a means of facilitating the interactions which must not go by default and of clarifying the areas in which individuals are able and expected to exercise discretion. In defining the role of a co-ordinator we are saying to him: these activities need to be co-ordinated and it is up to you to use your skill, your judgement and the authority you have been given to make sure that they are.

The great advantage of attempting to define relationships is that the definitions, and the process of arriving at them, can be used as a vehicle for building mutual understanding and consensus. Within the strategy of 'bilateral' implementation discussed below, the exercise of trying to arrive at an agreed set of role definitions provides a means for joint organisation design, for constructing a new pattern of interactions and a way of working together. Two things are needed for this task: a vocabulary and a process. By a vocabulary I mean a shared awareness of a set of options, such as the organisational models and role definitions given in Chapter 10. One of the great inhibitors to spelling out relationships in a matrix has been the oversimplified view that only two forms of influencing relations are possible: manager–subordinate and staff–line. It may well be that this, rather than any objective consideration, is the most common reason for creating project or product 'managers' without defined authority. Once there is a shared understanding of the options, constructive organisation building becomes possible.

The process for doing this seems to require some form of catalyst – a person or a task. Social analysis, the collaborative research method referred to in Chapter 10, is one example of such a process; here the catalyst is the social analyst, an outsider helping organisation members by questioning them to clarify the requirements of their collaboration. In a bilateral implementation process it might be the chairman of a working party or a consultant operating in the way described by Sheane in Chapter 13. Another example is the responsibility charting process which Sheane also describes. Here the chart itself provides the catalyst, by presenting the questions on which the team members have to come to some agreement.[1]

173

While it is not possible to generalise for all circumstances, and say that definition of role relations in a matrix is always desirable, it does seem to me that in most situations, short of the most committed 'task cultures', a participative attempt to construct an agreed framework of relationships can be a powerful antidote to dysfunctional conflict, a basis for confronting issues critical to effective collaboration and a means of avoiding confusion, buck-passing and delay.

Power

Two issues which are critical to the introduction and management of matrix structures are often avoided, because both of them have to do with the real power (as distinct from formal authority) of individuals and groups – a topic which, though of consuming interest to the members of most organisations, is not considered a legitimate subject for discussion. It smacks too much of 'internal politics', a field of endeavour which many engage in but few approve of. The first of these issues is the effect of introducing a matrix structure on the existing distribution of power – who will gain, who will lose? Managers' pre-occupation with this question can be one of the chief causes of the dysfunctional conflicts to which matrices seem to be subject. Fear of this often results in the approach mentioned earlier, of setting up horizontal roles (project or product managers) with defined responsibilities but no formal authority, a move which is thought to avoid the appearance of threatening established power positions. This, however, raises the second power issue: how can the people in these horizontal roles be given the real power to do their jobs?

Power, particularly in the first context, means more than the commonly accepted definition of one individual's ability to determine or influence the actions of others. It includes the ability to get and keep control of resources and to operate autonomously within boundaries under one's control, without outside interference and with full jurisdiction over certain areas of work. It may also include early access to information and the likelihood of being consulted about decisions outside one's own sphere of responsibility. Obviously these facilities are highly prized, and those who possess them tend to be well placed to defend them against outside threat.

Where does this power come from? The obvious answer is from formal authority, hierarchical rank and status. Two other sources are often added: professional expertise ('sapiential power') and personal relationships ('referent power'). Research, however, suggests two important determinants of organisational power which are quite distinct from these. One is control of information (Pettigrew, 1972, 1973). The other is the ability of individuals or groups to help the organisation cope with critical forms of uncertainty which face it (Crozier, 1964), particularly if those groups are in a central position and difficult to replace (Hinings et al., 1974). Because of these factors the introduction of matrix leads to changes in power distribution even if definitions of authority remain unaltered.

The strategic position of departmental managers in the vertical communication channels, by virtue of which they can control information flows, will be greatly diminished through the increase in lateral communications. With information widely disseminated it is no longer the control of information but the ability to use it that confers power – a shift from position power to expert power. Departmental boundaries and autonomy within them become more difficult to maintain. Through cross-functional operation cherished practices are thrown open to inspection by outsiders, and have to be justified. Simultaneously, the focus for dealing with important areas of uncertainty, whether of markets, technical problems or achievement of deadlines moves out of the specialist department into the interdepartmental team, strengthening the horizontal managers at the expense of the vertical.

What evidence there is of the effect of matrix organisation on power distribution all points to a weakening of the senior departmental or 'line' managers in relation to those with direct project or product involvement, both project managers and more junior staff in the departments.[2] Another unexpected result noted by Dalton et al. (1968) is an increase in the power of the head of the organisation. This seems to be a case of weakening the barons and strengthening the king (cf. Jay, 1967), the top man being the only remaining 'cross-over point', with senior managers being less able to censor the information that reaches him.

Given these effects of matrix organisation, it is not surprising that the existing senior managers of a functional structure should often be opposed to it, and should find themselves in conflict with project and product managers. But what is the answer? There may be many answers or none, certainly no obvious universal remedy. But two points can be made. One is that this effect on the existing power set-up emphasises the importance that attaches to a well thought out process of implementation, discussed in the next section. The other is that it is not enough just to define the new horizontal roles and their responsibilities. The jobs of the existing departmental heads need to be reviewed at the same time, both to create the correct balance between the two, and also to ensure that their stake in the new organisation and their contribution to it are not diminished.

The second issue concerns the power of project or product managers. This can be a problem even if they are given explicit authority, but it is much more so in its absence (cf. Wilemon and Gemmill, 1971). Two aspects have to be considered: the factors to be taken into account in selecting people for the job, and the way in which the role itself is set up. Important factors in selection are:

> personal qualities, such as tolerance of ambiguity, powers of persuasion and ability to establish good relationships;
>
> prior relationships, contacts and standing, and facility in knowing one's way around the organisation;
>
> ability to command colleagues' respect through technical competence and expertise.

These personal characteristics can be greatly assisted by the way in which the role itself is established and integrated:

> reporting relationships at an organisational level which confers the required status;
>
> communication systems which give the project manager access to key individuals in the organisation, so that people feel that collaboration with him can help them in their careers;
>
> financial resources, such as control over certain parts of project or product budgets. (In some project management systems the project manager negotiates allocation of the project budget with department heads.)

Unless a conscious attempt is made to equip project or product managers with some such specific sources of power, the 'responsibility without defined authority' approach is likely to prove ineffective.

Implementation

Not all matrix organisations are consciously implemented; some evolve naturally over time and are only recognised subsequently, as in Sheane's account of their development in ICI (Chapter 3). In these cases where a matrix system is the cumulative outcome of a succession of local changes, people sometimes wonder what all the fuss (about matrix management) is about. Jaques (1976, p. 259) indeed regards the concept of matrix organisation as simply a hold-all for the many lateral relationships which are bound to grow up in all bureaucratic organisations. The fact remains, however, that in at least some organisations, at a definable point in their history, a decision is taken to introduce a matrix.

In the second stage of our survey we asked about the way in which this had been done. The answers fell into four groups: in the first two, roughly equal in size, the structure had either evolved or had been implemented unilaterally by top management decision, while the third, rather larger group reported that implementation had been bilateral in consultation with the managers involved and sometimes their staff. The fourth group reported a mixture of these alternatives.[3] The only description of an implementation among our case studies is that in Scicon, which was conducted bilaterally and achieved considerable success. In contrast Argyris (1967) describes a matrix which was imposed unilaterally and was not very effective. I do not think that we can deduce from these two examples that bilateral implementation always succeeds and unilateral never. Indeed, among the respondents to our survey, those who described their implementation as unilateral reported a slightly higher success rate in achievement of objectives and slightly fewer serious problems than the bilateral group, though neither of the differences was statistically significant.[4]

There may, however, be a difference in the effectiveness of the two approaches when the objectives of the reorganisation are taken into account. In Chapter 8 I suggested that the method of implementation should be such as to make the achievement of the objectives more rather than less likely. Thus it seems sensible to suppose that where the aim is either to improve spontaneous co-ordination of the efforts of different groups, or a rise in adaptive potential through flexible and unprogrammed interaction, or even an increase in participation and commitment, then a bilateral, participative style of implementation may help to get these processes started by bringing the different parties involved together in working out the new structures. An imposed change on the other hand may, as in the case described by Argyris, generate hostility and suspicions which are carried over into the operation of the new system. The responses to our survey are compatible with this view. While again not reaching statistical significance, they suggest that a unilateral approach is better at achieving efficiency, control and accountability objectives than at improving co-ordination, adaptation or social effectiveness, while bilateral implementation seems to be slightly more successful with the second set of objectives than the first.

Argyris feels very strongly that a unilateral process of implementation generates conflict and tension in the organisation and defeats the aims of matrix management. He envisages an extended change process running through a number of phases of joint problem diagnosis and consultative organisation design which provides 'a living educational experience for individuals and groups on how to work together; on how to develop internal commitment among the members of the organisation, and how to reduce the unnecessary and destructive win-lose rivalries' (Argyris, 1967, p. 55).

The approach to implementation needs to be considered in the light of the changes in relative power which a matrix is likely to bring about. Where a new structure is imposed, this very fact automatically constitutes a reduction in the power of all those who are neither privy to the reorganisation nor its direct beneficiaries. Hence the possible effects of the new structure are compounded by the manner of its implementation. On the other hand, involving managers in the decisions underlying a reorganisation is a way of enhancing their power, and with it their sense of security. To the extent that they share control over the change process they are in a position to affect those aspects of the structure which would otherwise involve them in the more harmful forms of conflict.

Training and organisation development [5]

A need for training and development can arise in the matrix both at the individual level and at the level of the teams which are often one of its most important features. For the individual there is the need to understand the structure and demands of a more complex environment, both within the organisation and around it, and for greater insight into the concerns, assumptions and outlooks of other team members belonging to different functional groups and backgrounds. He may also need to develop social skills of a different order from those required in working with colleagues who share his own frame of reference – Bridger calls these 'consultative skills'.

The problem for the team is not only, as is often supposed, one of internal group dynamics and interactive skills, but of learning how to cope with the intensely political context in which it has to operate, and of discovering how to integrate the contributions of the interdependent but separate home bases of its members to the common task. Indeed the word 'team' itself can, as Johnston has pointed out, be very misleading, by its analogy with sporting teams.[6] Where the latter are self-contained and in full control of their own boundaries and resources, the matrix team has to learn to operate as a more or less open system (more in the co-ordination matrix and overlay, less in the secondment matrix) whose effectiveness is related to its ability to manage relationships outside its own boundaries.

Attempts to meet these development needs of groups and individuals can take the form either of in-company organisation development activities or of training experiences which bring together members from a number of organisations in a more neutral setting. Typical of the former approach is the team-building workshop described by Sheane in the next chapter, in which the members of an actual or potential team are helped by a consultant to examine both the problems which they face and the appropriateness of their normal ways of dealing with them. Bridger feels, however, that in many organisations the existing culture and relationships would make it difficult for the problems and conflicts of interdependence to be faced openly and insightfully enough for an internal workshop like the one described by Sheane to have much chance of success initially. Under these circumstances external training can provide a valuable input. Two approaches are possible. The first, knowledge-based and conceptual, tries to increase understanding of the organisation's relations with its environment, its internal structure and the interdependence of its parts. The other approach is more experiential, and focuses on the processes of interaction in the face-to-face group and the motives, conscious and unconscious, which underlie them. In Bridger's opinion neither of these approaches is sufficient by itself; what is needed is a combination of the two.

What might a workshop which combines these two approaches set out to achieve in more specific terms, and how would it go about it? In one example, given by Bridger, the objectives include increasing both the understanding and the skills of the participants. Increased understanding of the issues which confront them in their own organisations, of the organisation's interaction – as an open system – with its environment, of the relations between its tasks, its technologies, its people, its operational sub-systems and its organisational options. Increased understanding also of the group processes that operate between members working together in a face-to-face team, and of the differences in the perceptions of people from different cultures (whether American and British, or marketing and research). Increased skills in working in a 'consultative' mode, i.e. with people from a different occupational group in a collateral relationship, skills in representing one's group in relations with other groups, skills in building, relinquishing and changing organisations (i.e. patterns of relationships). Above all such a workshop aims to build up learning skills; it is a 'learning to learn' experience – learning, in the workshop, to learn in the back-home situation. The methods used to achieve these objectives include preparatory work of 'personal mapping', working together in learning groups and consulting groups, intergroup communication, institution building and conceptual inputs through teaching seminars.

The *preparatory work* which participants are invited to undertake before the workshop is an open-ended exercise in mapping their own current situation to discover the personal, occupational, organisational and social factors which are helping to shape their activities, their attitudes and their scope. At the same time they are invited to consider the problems facing them in their work and are offered criteria for selecting significant issues to work on in the learning and consulting groups.

The *learning groups*, which consist of up to eight participants working with two staff members, are given two parallel tasks, one concerned with content, the other with process. The content task is to study issues arising from the members' organisational concerns and to come up with ideas and principles which can help to resolve these issues, while the process task consists in exploring the actual interactions and relationships which arise during the work on these issues.

The *consulting groups* of three or four members focus on members' concrete problems and projects: members try to assist each other towards finding solutions of their problems and, in the process, develop their own consultative skills.

Intergroup communications provide experience of the problems of communication between groups, and their effects on the groups and their representatives.

Experience in '*institution building*' is provided by treating the workshop itself as a temporary organisation or institution, which can be changed by the members in response to emerging needs.

Seminars include conceptual inputs on such subjects as the dynamics of the workgroup, process consultation, and organisation design and development.

Such external workshops can be appropriate for organisations whose culture is not sufficiently 'open' for issues vital to the functioning of a matrix to be tackled head-on. By the 'learning to learn' opportunities which they provide they can build a foundation for later internal development and training activities, including the basic organisation-building, role-defining task discussed in an earlier section.

Support systems

Organisation structures and management systems ought to be compatible

180

and mutually reinforcing; this is obvious and non-controversial. But, as Hopwood notes in Chapter 14, structures and systems are not always sponsored or designed by the same people, and it is likely that many matrix organisations are being frustrated by having to struggle along under the handicap of inappropriate systems which are more of a hindrance than a support. There are two main areas where this danger arises. One is the field of management information and control systems, usually the preserve of the accountant and the systems analyst. This area is examined in some depth by Hopwood below. The other is the sphere of personnel management. While it is possible to point out the problems that can arise in the latter sphere and while the answers to most of them do not even seem particularly difficult, I am afraid that very little hard information on company practices has come my way and I am reduced to speculating about the subject. The obvious areas in which practice ought to take account of the existence of a matrix structure, seem to be selection, management development, appraisal, job evaluation and manpower planning.

Selection

There is good reason to believe that individuals with a low tolerance of ambiguity may find certain matrix structures difficult to work in, though I do not know of any proper evidence of this.[7] Psychological tests for tolerance of ambiguity do exist, but I have not heard of such tests being incorporated in any selection procedure.[8] One would imagine that this factor would be particularly important in the selection of project or product managers, and even more so if the company practises a policy of not defining these roles in any but the most general terms. Some of the other factors which are important in the selection of project managers were mentioned earlier (p. 176).[9]

Management development

Matrix management provides both a challenge and an opportunity to the management development adviser. The challenge is to devise programmes of training and personnel development which will prepare specialists to take on the very wide responsibilities of a manager or co-ordinator in a matrix. The opportunity is the existence of a set of middle management roles with general management responsibilities, those of the project or product managers, which can be used to give early managerial experience to individuals judged to be 'high fliers'.

Appraisal

This can cause problems if the system does not recognise the existence of dual accountability. According to one of Gunz and Pearson's examples (p. 36) it can also cause trouble if it does. My own view, in spite of the case they quote, is that in overlay and some secondment matrices both managers need to be involved in the appraisal process.

Job evaluation

Again it seems obvious that increased responsibility must be recognised by the evaluation system and result in increased pay, but companies have been heard of where project manager appointments 'do not count' for job evaluation purposes. Presumably increasing unionisation of specialists and middle managers will put an end to such practices. The precise evaluation of these jobs will of course depend on the way in which they are defined and on the criteria employed by the job evaluation system in use.

Manpower planning

The models used by manpower planners will have to be revised when a matrix organisation is introduced, to reflect the change in promotion prospects, the increase in requirements for managerial staff and the effect of these changes on middle management turnover. In a project management matrix there may also be a need for special models to represent the build-up, run-down, phase-out and redeployment of manpower on individual projects.

Summary

Approaches which seem to be relevant to the problems of balance, conflict, individual adaptation and administrative cost are sketched out and then considered individually. Though flexibility in operating a matrix is often thought to depend on some ambiguity of roles, a decision to define role relationships need not have the effect of restricting initiative, and the process for doing so can play an important part in participative implementation. The effect of a matrix organisation on the distribution of power ought to be considered in deciding how to implement the structure and in establishing new roles. The approach to these questions may affect the ability to achieve the objectives of matrix working. Development of the additional understanding and skills required by participants in a matrix organisation can be assisted by in-company organisation development activities and by external training experiences. It is important that management support systems should reinforce rather than hamper the matrix structure: the issue of compatibility arises both in management information and control systems and in various areas of personnel management.

Notes

[1] Another method of defining responsibilities in a matrix is proposed by Schemkes (1974). This involves the use of an elaborate questioning approach, reminiscent of the check-lists used in method study, with questions like Who? When? How? etc.

[2] Dalton, Barnes and Zaleznik (1968) provide the only study I know of to look explicitly at the effect of a reorganisation with features of matrix management on the internal distribution of power. Supporting evidence comes from Kingdon (1973, p.137), Hill (1974) and Steiner and Ryan (1968, p. 31).

[3] The numbers were – evolved: 16; unilateral: 15; bilateral: 26; mixed: 10.

[4] These differences may be related to the fact that the bilateral implementations tended on average to be a little more recent than the unilateral. Also our data are not about facts but about perceptions – unilateral managers might conceivably be more ready to perceive success and less willing to admit problems. It is hoped to publish the full results of the survey later.

[5] I am indebted for much of the content of this section to some very helpful discussions with Harold Bridger of the Tavistock Institute of Human Relations, based on his experience of conducting workshops and training groups for members of matrix and other complex organisations.

[6] Talk by Arthur Johnston to the Organisation Development Network, 1976.

[7] One organisation which operates an overlay matrix finds that newcomers either like the system or strongly dislike it and that those who dislike it tend to leave after a short time.

[8] One of the second order factors in the 16PF test, which is widely used in selection, probably provides a good approximation to tolerance of ambiguity. This is factor IV – 'Subduedness v Independence'.

[9] I gather that a research project on the characteristics of successful project managers is currently under consideration in the UK.

13   An organisation development approach

DEREK SHEANE

This approach to improving matrix operation is best dealt with by describing in some detail the work done in an actual situation. This will be followed by an articulation of some of the principles applied. The situation to be described is disguised for reasons of confidentiality, and the principles drawn out at the end not only reflect this particular case but can be assumed to be a generalisation of the approach used by many managers and 'change' resources in ICI.

Setting up an OD activity

The general manager of a fairly substantial (£10 million sales) business expressed some initial concern about the problems he saw in running a business effectively in the commercial and social world of 1975 onwards. His initial concern focused on such items as the needs to improve manpower productivity, decrease the level of stocks, simplify the organisation structure and get the different functions (production, sales, research, etc.) working more effectively. Central to the general manager's concern was the question of matrix organisation. Up to now this business was run using traditional hierarchies but there was a considerable ongoing natural development of 'cross-functional informal' relations. Indeed one question was: should these informal processes be made 'legal' and should the business adopt matrix organisation as a key method of organising, or should a predominantly functional or departmental approach prevail? The general manager's response was to initiate a study team to consider how the business should be run and this study team, in terms of composition, reflected the type of cross-functional teams typical of ICI matrix organisation: production, marketing, research and planning inputs.

His concern was initially expressed to his local personnel manager who was not only involved in the routine work of the personnel function but was also highly concerned about the need to initiate development work that would contribute to longer term health, as well as short term performance. At much the same time the general manager had read some material prepared by the author for ICI as a whole on the problems associated with running and improving complex and matrix type organisations as opposed to simple hierarchical ones. This led to an informal meeting initiated by the personnel manager where he and the general manager outlined the problem situation, this time in more detail, and in these categories: business, organisation, behaviour and systems. Examples of issues within the categories were: what business policy is to be adopted opposite distributors on stock levels? Is it possible to use a cross functional business team to improve overall performance? How can we get people to behave differently – to handle conflict more skilfully? What can we do about the ineffectiveness of the call-up system? From the point of view of matrix working some of the key questions in their minds were: what type of cross-functional team would be best? What mechanisms and processes are best suited for developing business policy and securing implementation? If we go wholeheartedly for a matrix system how can we make sure changes in understanding and behaviour – for example a more direct and open conflict resolving style – can be developed? These are the types of difficulty that are, of course, experienced in simple hierarchical organisations but become much more dysfunctional in a matrix situation.

The author, in the role of external consultant, mostly listened and asked questions in order to get a good feel for what type of approach might start and maintain an ongoing improvement process. Towards the end of the meeting he made some points, mostly obvious, but nevertheless critical.

1   The problems were interdependent: it seemed impossible to isolate, say, the question of matrix working across functions and treat it as a self-contained issue.

2   The situation was a microcosm of certain well-known but 'hard to solve' problems, for example, high vested interest in maintaining the *status quo*, and a lack of a sense of direction in terms of 'where the business was going'. These underlying problems would have to be confronted directly if progress were to be made. This would involve pain and discomfort for individuals.

3   Whatever solutions this business came up with would have to be sold to or somehow integrated into the larger ICI system. This again was likely to be uncomfortable for both parties. However, this was typical of most situations nowadays as it is part of the wider social and economic situation: namely that many individuals and groups had 'out of control' feelings. Therefore, the question of the amount

of energy individuals in this team had for facing up to inertia in the total system would have to be faced.

4 So far we had only two definitions of the problems, the general manager's and the personnel manager's. Other perceptions would be needed as it is characteristic of matrix organisation that agreeing a common appreciation of the issues and problems is very difficult, but if agreed probably the critical factor in starting to make progress.

At the end of the meeting a summary of the discussion was put together and it was agreed that the personnel manager and the external consultant should produce some ideas on how this problem might be approached and present them to a wider audience, namely the senior management of the business that needed improvement. The personnel manager was established as the internal consultant, a role he was to develop further as time went on.

The external consultant identified three alternative approaches which he felt would give enough real choice and at the same time give direct guidance on how the situation could be tackled. The options were:

1 A formal organisation study done as action research: the consultant to collect key data individually from key participants and feed it back to the business team as a group, leaving them to plan actions.

2 The external consultant to attend a cross section of activities, business meetings, cross-functional work groups, etc., and make a series of interventions aimed at improving effectiveness. This approach, widely used in ICI and elsewhere, is usually known as 'process consultation' where the word 'process' indicates that the consultant focuses on issues like decision making and conflict management – the processes of management – as opposed to the content: for example, rather than suggesting what level of stock would be appropriate, the consultant would help the team develop a good process for handling the issue.

3 An intensive workshop, where all key participants gathered together and stayed together until they came up with solutions. The consultants would help keep this process on the rails and intervene informally and on an 'as needed' basis. Their contribution could be either process intervention or recommendation on the content if within their competence.

Having thought these options out the external consultant attended the larger meeting of the general manager plus three functional heads. Instead

186

of presenting the options at the start of the meeting the consultants agreed with the general manager that it would be highly desirable to take the group through the process he himself had been through at an earlier meeting: i.e. moving from a 'mess' towards a clearer definition of the issues and problems. It was also pointed out that intervention and change was now already under way. For example, this meeting, if handled correctly, would bring out the different perceptions and conflicts.

This indeed was what happened. There was no shortage of views, and of course, great variation not only in problem definition but in solutions. Also, the behaviour problems encountered in matrix working were on display, as the meeting was a microcosm. For example, differences between production and sales were swept under the carpet, the general manager did not act as a third party but favoured his 'old function', and there was confusion about roles and responsibilities. (I talk of these as if they were matrix problems only – it is, of course, the case that these are sometimes general organisational problems. My impression is, however, that they are especially chronic in complex forms of organisation of which a matrix is one type.) Later in the meeting the consultants and the general manager, who was now helping considerably by keeping the group focused on the problems and by emphasising the need to face up to differences, asked the group to conclude what their views of the fundamental problems were and asked if their impression as outsiders that individuals saw different problems, defined the same issue differently and were proposing conflicting solutions, was valid. Furthermore, did they accept that this in itself was a significant hurdle to be overcome? Both ideas were accepted after further discussion and clarification.

The general manager asked the external consultant to outline his approach to tackling the whole situation, including the process of involving others connected with the business but not at the meeting. The three options were outlined, but no preference was stated by the consultant. The only comment was that it was important for them to choose their approach on a basis of real consensus, as whatever method was chosen would involve a major time, financial and personal investment. It was better to argue it out before rather than after the event. After a lengthy discussion which exposed varying degrees of commitment to the different approaches, the third option (the intensive workshop) was chosen. At the end of the meeting it was agreed that the general manager, after taking advice from colleagues on the details of who should be involved, would send out a letter outlining the approach.

After the letter went out announcing the details to the rest of the business, the personnel manager in his role as internal consultant initiated a session with the potential total business group (ten people) to help prepare them for the workshop, which was going to consist of at least three days' work including evenings, plus a fourth day if required. This meeting was attended by the external consultant and the two most senior managers (production and sales), who were at the previous meeting. It was decided that the general manager should delegate the workshop task to these two and take no further part until after the workshop.

This meeting was a key event as there were issues that had to be dealt with before any real commitment from the larger group could be obtained. Their concerns, which had to be more or less teased out with patient discussion, were: Why were we not consulted about the workshop? What is there to spend three days on? Is this going to be a session on personality and interpersonal behaviour? How did you decide who to include and who should be left out? These questions were dealt with factually, the senior management more or less reporting on the last meeting and asking the consultants to describe the very earliest stages when only the general manager was involved. The session went on for about two hours when again the process of the meeting provided data rich in content and quality about the problems the business had. The main task for the consultant in this session was to attempt to establish realistic expectations from the workshop, and build confidence that, if people worked well together and straightforwardly, then good plans for improving the business would emerge. As always, not much was achieved in this regard and, when the workshop began on a Monday evening, three weeks after this meeting and two months after the initial meeting with the general manager, commitment was varied, to say the least. There was even a general atmosphere of apprehension and doubt about the wisdom of the approach.

The workshop

The first evening was spent developing aims and expectations. Each team member was asked individually to consider two questions: firstly, what do I expect personally, and hope for from this event? Secondly, what would be a good output from the workshop for the group as a whole?

This was a 'new' team in the sense that, although they all knew each other, they had not worked together as a team before. Also, everyone except one of the senior managers (who was the product/market manager) was a part-time member of this business team. As it was part of a divisional matrix organisation, each individual had one or more other significant business or functional commitments: for example, the production manager managed a system that supplied products to two other businesses, and that competed for allocations of capital and resources from the same divisional fund. For these reasons the questions above were aimed at obtaining awareness and practice in thinking about the conflicts in personal role brought about by dual membership: being a member of the basic reference group ('my function') and also contributing to a cross-functional group with goals above and beyond but in conflict with functional goals. A critical and tense example was the need for production efficiency clashing with the need to be market oriented and serve a dispersed group of small end-user customers – not an easy task.

After the workshop had been considered in terms of these questions, the thoughts and needs of the individuals were built up into a group product. The group needed a great deal of 'process help' but finally got there and felt well enough satisfied. This was the first time they had accomplished a group task and the opportunity was taken to review the quality of group work. In accordance with developing the habit of productive self criticism and open discussion the team discussed the questions: How did we perform? What was useful, and what could be improved?

Expectations having been brought into reasonable congruence, the rest of the evening was devoted to planning the three days. Without labouring details the members opted for simplicity: Tuesday would be defining and agreeing basic problems; Wednesday would be broad strategy; Thursday would be action planning for back-home, and the evenings and Friday would be left free to create spaces for discussion of the need that might arise on the way.

This plan was followed as a basic model but with many diversions and going back over previous areas. As the consultants had more to contribute to, and greater interest in, the 'process' rather than the substance of the meeting, division of labour and roles naturally occurred. The business team were in charge and felt full responsibility for outcomes but expected the consultants to help identify blockages to effective matrix working and to provide insight and 'ways of looking at the situation' based on wider experience of other ICI situations. Rather than attempt to give a blow-by-blow account, I have listed some of the inputs or educational assistance given on an 'as needed' basis.

Tuesday matrix organisation: the basic framework for choosing different forms based on Jay Galbraith's 'Designing complex organisations' (Galbraith, 1973).

Wednesday group working: using examples of actual group behaviour as material for developing insight into how workgroups avoid problems through various mechanisms of defence, based on Bion's 'Experiences in groups' with dependency, fight, flight and other processes being interpreted (Bion, 1961).

Thursday intergroup difficulties: descriptions of the various degrees of health or disorder that can exist between groups: functions, specialisms and so on. Advice on the different mechanisms and processes for resolving conflicts.

This and other types of educational help were inserted as the workshop progressed with care taken to ensure relevance and comprehension. A typical educational intervention would consist of about fifteen minutes of presented material followed up by at least the same or more discussion time. A significant exception was the preliminary session on basic forms of matrix organisation which lasted two hours with roughly equal amounts of input and discussion.

As the workshop progressed the question of action planning and making an impact on the larger system started to become paramount in individual minds. The fact of working together on real problems, regular review of how the group was working (whether as a totality or as subgroups) and the educational and process help given had started to produce an effective working unit. But, in a complex organisation, the real barrier to improvement and higher performance and self-fulfilment is often the refusal of surrounding systems to tolerate differences and innovations, whether we are talking in terms of new organisation structures, systems, processes or behaviours. Consequently, as time went on, more effort had to be directed towards developing a good influence process to ensure implementation and follow through. In practical terms it meant thinking through how to sell the ideas to bosses, colleagues and subordinates. It also meant generating awareness that one group's initiative could be another's loss of power, capital, freedom of action, etc., and that any proposal would have to recognise the needs of a larger system. Out of this theme emerged a series of first steps that would enable the introduction of new ideas and concepts into the power system as well as the normal daily work of the organisation. For example, a proposal to simplify the planning system implied the loss of a job for an individual. The plan to decrease stock levels and be more efficient meant the marketing people had to develop a new attitude to customers, a tougher one, in fact. These two examples show how, unless there is real commitment to optimising the total business and making sacrifices somewhere, and an agreement to keep reviewing progress, the likelihood of increased performance is very low.

By Thursday it was clear that the work was not yet finished: in particular, there was a need to clarify roles and responsibilities within the business group. So the workshop extended to Friday; this could be seen as Parkinson's Law in operation but, from the atmosphere, it was easy to see that energy was running at a high level and that there was a willingness to solve these questions, although they were threatening and anxiety-provoking.

Friday's session implied working through a responsibility chart, an essential tool for clarifying and re-negotiating roles and responsibility. Those interested in the detail of the technique should see the final chapter in Galbraith's book mentioned before.[1] The outcome of the session was a chart showing how different functions and individuals would contribute to decisions. For example, if the issue was 'manpower allocation to new product development', then the chart would show who was responsible, who needed to be consulted, who needed only to be informed and who (if any) had a veto power. The chart (Fig. 13.1) is totally unlike a formal hierarchical chart and this is best shown by asking what would happen if an outsider said 'show me your organisation chart'. A team member would respond by saying 'for which decision?', indicating he had a matrix where the axes were issues and people or functions and the squares showed the type of involvement.

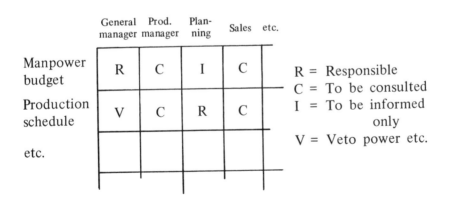

	General manager	Prod. manager	Plan-ning	Sales	etc.
Manpower budget	R	C	I	C	
Production schedule	V	C	R	C	
etc.					

R = Responsible
C = To be consulted
I = To be informed only
V = Veto power etc.

Fig. 13.1   Responsibility chart

This chart was also used for discussing how the group would operate in future: should it be by consensus? Should the whole group meet? Agreement on matters such as these was rounded off by a standard setting session where the group agreed a set of behavioural ground rules for all future meetings: for example, concerns to be aired in the meeting not in the car going home; conflict to be confronted not avoided; to be informal not bureaucratic, etc. The workshop ended by summarising back-home action plans and setting a date for the first review meeting.

Follow up

The organisation development approach described so far illustrates the usual problems in getting action in the midst of complexity, vested interest and uncertainty. By the end of the workshop the initial goals in improving matrix working had been achieved. Important achievements as far as the team was concerned were:

(a)   clarity about goals and aims;
(b)   cohesion on the philosophy of running the business;
(c)   agreement about the nature of roles and how the functional and specialist inputs would be built into the business and organisation plans.

The consultants considered these changes worthwhile as:

1   The team had a common language for discussing problems.
2   They had a much higher level of awareness of the skills needed to manage matrix and had a fair amount of practice/review/more practice over the four days.

The general manager, on getting individual feedback on how the workshop had gone, was impressed by the amount of energy released and the intentions expressed. He now felt better placed to handle his own larger power system.

Although all of this seems good it is important to underscore the need to intervene to build in an ongoing improvement process. I feel that a repetitive 'going away to sort things out' process means only one thing: that improvement is still a one-off exercise and the learning and energy generated are not being used at the place of work. Consequently, several follow-up interventions were made to sustain progress: periodic review-of-progress meetings on an 'as needed' basis; action by the general manager to 'protect' and develop the work the group was doing; mild harassment by the internal consultant to remind people of commitments when energy was flagging. With this kind of measures the changes are incrementally built into the system and the gains realised. As far as hard evidence of success is concerned, movement towards the quantified targets on stock and other efficiencies is being achieved. Evidence on 'softer' items like behaviour and co-operation is harder to indicate. Although there is widespread expressed improvement I know of no way, except a detailed third party 'before and after' study, to ascertain this. Of course, to be absolutely rigorous there is no way of answering the question: what would have happened if there had been no OD intervention? My own assessment, based on my view of the utility of good OD activities is: they speed up what is needed and may have happened anyway, and the experience leaves behind some 'good' organisational habits such as being clear about purposes or being more skilled in group behaviour.

Conclusion and summary

If this is a typical example and is held up as a success, then what are the principles underlying its design and implementation? The reader can assume that the principles outlined here can also be seen as emerging from examining failure situations, which, whilst fortunately infrequent, do occur. In no order of importance the principles are as follows:

1   A felt need was picked up skilfully by an internal resource, the personnel manager.
2   The personnel man was confident and personally secure enough to introduce the general manager to an organisation development approach and an external third party.
3   There was an emphasis on diagnosis in the early stages and a willingness to go through a somewhat repetitive process to achieve a 'common appreciation' of problems by different levels and functions.
4   Two types of issues were worked on simultaneously: task and process.
5   The barrier of the potential resistance to change by the total system was thoroughly worked through and in the open.
6   The external consultant had seen other similar situations and could at least demonstrate understanding and thereby not be rejected early on: his experience was both theoretical and practical.
7   The power group chose the basic method of intervention and

were therefore ready to handle the output from the activity.

8   The workshop as an experience was enjoyed: it was hard work, but informal; on occasions it was humorous and at all times it was real.

9   There was a process of ongoing review – stopping to get feedback on how things were progressing. This mechanism was instrumental in putting issues on the table openly and kept an automatic check on general direction and usefulness of the work.

10   Everyone had their say. Work was structured by sub-groups, by pairs and in other ways so that productivity could be paralleled by meaningful individual participation.

11   There was a limited amount of structure as well as flexibility. Using resources on an 'as needed' basis was established.

There are other lessons to be learned but these strike me as key points if an organisation development approach is to succeed. Also, although matrix organisations have many distinctive characteristics, the process of changing them and the principles to be applied in managing change are very similar. Finally, the participants would not recognise that they had used an OD approach. Their description of what happened would be something like: 'we went away for a few days to give the business and ourselves an overhaul and we are now following up our plans'.

This type of reaction and attitude is what I would think should characterise a healthy customer response to the first attempt to improve matrix working by using an organisation development approach.

Note

[1]   See Galbraith (1973) and a more detailed description of the method in Melcher (1967), reprinted in Cleland and King (1969).

14 The design of information systems for matrix organisations

ANTHONY G. HOPWOOD

The design of appropriate information and control systems should be a crucial part of any attempt to develop and implement a matrix organisation. Matrix organisations, like any other management structures, require information to function effectively. Planning procedures, information for decision making and systematic ways of monitoring and controlling performance are an essential part of the management process. Indeed for matrix organisations the need may be all the greater. For the matrix approach is invariably implanted within an existing organisation and there are usually innumerable forces that have the potential to frustrate its success. There is invariably a real need for carefully designed information and control systems that explicitly aim to support the matrix rather than neutralise or even destroy its potential. Yet, on so many occasions, the seemingly obvious need for appropriate information systems is not given the emphasis that it deserves, and matrix structures are left to exist alongside information and control systems orientated towards a previous functional management structure.

In part this reflects a much more general management problem. The complexities facing modern organisations, and their increasing size, have resulted in a fragmentation of not only the actual management task but also the evolving body of management expertise and knowledge. Once started, this fragmentation process develops its own momentum. Technical and functional boundaries become jealously guarded, and, with time, as different management groups respond to differing circumstances and the pressures of their increasingly separate environments, leads and lags, if not stark inconsistencies in managerial practices, become increasingly apparent.

Of course, matrix organisations are intended to provide one way of overcoming these very problems, but the same factors which provide the impetus for their development can also constrain their effective implementation. In no area is this more readily observable than in the design and use of complementary information systems. The technical concerns of the designers of information and control systems, and the behavioural and organisational assumptions that they implicitly make, can easily appear to be, and frequently are, very distant from the concerns and assumptions of those organisational designers who have fostered matrix approaches.

An illustrative example

That the fragmentation of expertise and efforts certainly has its attendant costs is illustrated by the endeavours of one major company that was experiencing tremendous difficulties in co-ordinating a complex set of highly interrelated activities. Recognising that co-ordination required both an appropriate organisation and more effective information systems, senior managers in the company proceeded to establish two high-level task forces in these areas. Both groups produced what appeared to be eminently reasonable recommendations and, with comparatively few changes, these were speedily implemented. But no sooner had this been done than the underlying problems got even worse. For whilst the organisational group had recommended the use of project and product managers to strengthen the co-ordinating role of lateral relationships in the company, the emphasis that the information group placed on clarifying existing responsibilities and producing more comprehensive departmental performance information did little either to highlight the troublesome interdepartmental problems or to help with their solution. Indeed, with the new reports providing more detailed and comprehensive information on the activities within departments, existing departmental loyalties and the power of the vertical patterns of influence were strengthened. As each manager strove to improve the reported results of his own unit, conflict between the units intensified and co-ordination became ever more difficult.

The recommendations of the information group were firmly based on the doctrines of individual responsibility and accountability that are so influential in accounting thought and practice. The group had thought that the previous co-ordination problems were a result of the fact that responsibility for some crucial areas of management activity was far too diffused. From their point of view it appeared that many managers could not be held responsible for all the consequences of their own actions. Hence they had sought to extend the degree of individual accountability and to do this they recognised that an improved information system was an essential requirement. With more comprehensive departmental information systems, they thought that it would be possible to motivate individual managers to achieve specific objectives which were in the interests of the company as a whole and for which they could be held personally responsible.

Within the context of their own separate bodies of expertise, both sets of recommendations had looked reasonable enough, but their orientations and aims conflicted. One had focused on the activities of separate units and their aggregation through the management hierarchy whilst the other had focused on the relationships between units. By making those differences explicit, their combined implementation intensified rather than reduced the difficulties of co-ordinating the company's complex activities. Rather than supporting the move towards more influential lateral relationships, the new and highly visible information and control system helped to undermine both their affective and analytical bases. It strengthened existing departmental loyalties and provided little or no information to improve the quality of interdepartmental decision making.

If, however, the dominant concern is with the effectiveness of the overall enterprise, both organisational structures and information and control systems need to be designed so that they reinforce each other's concerns with the wider ends rather than dissipate each other's potential in the creation of additional conflict. Both, in other words, need to be seen as part of an organisational design endeavour. For, as Galbraith (1973, pp. 4–6) has pointed out, both are means by which enterprises strive to cope with the uncertainties and complexities of the task which they manage. Although the approach in practice may depend on the ends to be fulfilled, the overall need is present whether the aim is to achieve better co-ordination, facilitate organisational responsiveness, promote participation and autonomy or emphasise the significance of organisational outcomes rather than just internal processes.

Some common problems

Given that there is a frequent tendency to maintain traditional information and control systems alongside a matrix organisation, or only to modify them marginally to deal with the most apparent inconsistencies, it is important to consider the nature of the problems that this can create. Many management information and control systems, and particularly responsibility accounting systems which are so influential in practice, concentrate on reporting the consequences of activities within particular sub-units of an enterprise and on the aggregation of the information vertically through the enterprise. In so doing they reflect not only the need to compile the aggregate information on the enterprise that is required by law, but also the evolving role that such information systems have come to play in facilitating the exercise of authority downwards through the enterprise. In contrast, however, matrix organisation structures frequently aim to facilitiate inter-unit co-ordination through a greater emphasis on lateral decision processes.

In reflecting vertically oriented conceptions of organisational control, such information systems can, if used rigidly, encourage managers to focus too narrowly on the operations internal to their own units to the detriment of their relationships with other interacting units. The structure of the information system can, in other words, itself encourage a fragmentation of managerial perceptions, and as this occurs, the attendant rivalries and conflicts can endanger the very lateral relationships that a matrix structure aims to facilitate.

The dominant role that so many of today's information systems play in enforcing the influence of a vertically oriented power structure also means that they emphasise the production of information for the evaluation and control of subordinates rather than providing the information which is needed for decision making. This is a particularly important problem when traditional information system designs are implemented within matrix organisations. For whilst the need for decision information within the functional management units of the matrix remains, the project managers have additional needs for decision information that can facilitate co-ordination and integration and more generally contribute to the overall decentralisation of power and influence.

The above factors suggest that functionally oriented information and control systems have the potential to frustrate the aims of a matrix organisation. Similarly, however, the implementation of an integrative information and control system without a complementary change in the management structure can also be of no avail. If the underlying objectives of the matrix approach are to be realised, the management structure and the information and control systems must be mutually reinforcing.

Towards a framework for information systems design

One way of looking at both the problems and the potential of designing information and control systems for matrix organisations is to consider two fundamental dimensions underlying system design. The first dimension relates to whether the system emphasises decision or control information. Information for decision making is oriented towards facilitating the

198

exercise of self control by managers, whilst information for control is oriented towards control by others, especially by hierarchical superiors. The second dimension relates to whether the information system is oriented towards facilitating the management of functional units, with an emphasis on the utilisation of resources, or towards the overall project or programme of activity to which these contribute, with an emphasis on end results. The two dimensions and the resultant information system categories are illustrated in Fig. 14.1.

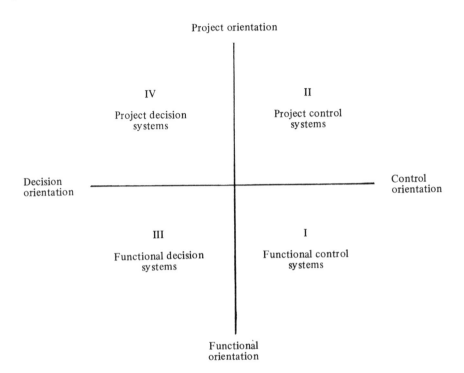

Fig. 14.1   A classification of information systems

Type I functional control systems are for monitoring the performance of functional management units. Many, if not most, of today's responsibility accounting systems, budgetary control systems and financial reporting procedures are of this type. Reporting on the utilisation of resources and the associated financial expenditures in areas of functional management responsibility, consideration is generally given to the relationship between financial inputs and outputs only at higher levels of the organisational hierarchy, and often then on a multi-product or project basis. Project control systems of the type II variety, on the other hand, are

199

explicitly oriented towards monitoring the performance of the separate projects and programmes of the enterprise. In contrast with the primary input orientation of the type I systems, type II systems are goal oriented, aiming to reflect the relationship between inputs and outputs at all relevant organisational levels on an ongoing basis.

Information systems for the financial control of projects report on project costs and revenues irrespective of the functional management units in which these have been incurred or realised (Cleland and King, 1975). Engineering, production, distribution and marketing costs are, for instance, related to the anticipated or realised revenues of particular projects. Similar procedures are used for monitoring the performance of contracts in sectors such as the construction and heavy engineering industries. In both of these cases the collection of the information in this way can be used to facilitate the monitoring of progress against time budgets.

The ability to express project outcomes in financial terms eases the design of type II control systems. But in the public sector, and also in some increasingly important areas of private sector activity, it is often impossible to express outcomes in financial terms. Nevertheless, some progress has been made on the design and implementation of related programme planning and budgeting systems (Anthony and Herzlinger, 1975). Extending beyond traditional departmental boundaries, such systems aim to relate all expenditures (and other indicators of resource utilisation) on particular programmes of activity to both quantitative and qualitative indices of programme performance. In local government, for instance, attempts have been made to relate expenditures on community health activities incurred within public health, welfare, education, recreation and sanitation departments to an array of indicators of community health including the incidence of diseases, mortality rate and hospitalisation statistics. Control systems of this type are undoubtedly in their early days. Numerous conceptual and practical problems remain to be solved. But this is not to say that they are not useful. If used with appropriate care they can greatly assist the management of complex and important activities.

The basic aim of types III and IV information systems should be to improve the quality of decisions made by individual managers of the enterprise. The type III systems aim to do this for the managers of functional units whilst the type IV systems strive to facilitate the activities of the project managers who are concerned with overall co-ordination and integration. In both cases the information should focus on those aspects of the job where the manager has discretion to act. If the flow of raw

200

materials, for instance, is beyond the control of a particular manager, information that aims to assist in their control is of no practical value to him. But where the flow is within a manager's discretion, information on the quantity, quality and timing of the flows and associated stocks, and more importantly, on the factors that influence these, is of potential value.

The fact that information systems of this type should reflect the nature of the job that they are aimed at assisting makes it difficult to generalise about the nature and scope of functional decision systems in particular. Different jobs have different information requirements and different management styles may also influence the demand for and use of information. Given these problems it should not be surprising that at this stage we know little about the general principles that might underlie such information systems. But this very ignorance is at least suggestive of the way in which they should be designed.

The basic question that should guide the design of functional decision systems should always be: what information do *you* need to do a better job? Any presumption that *we* know better should be viewed with some caution. Managers may have difficulty responding but the fact that so many do design and operate their own personal information systems suggests that the problem is mainly one of asking the right questions. Given that these systems aim to assist individual managers, their design should, as we consider again later, be of a participative nature, actively drawing on the resources and ideas of both information specialists and the ultimate users.

Project decision systems of the type IV variety should aim to facilitate the activities of the project managers who are concerned with overall project co-ordination and control. Although project managers should also be involved in the design of these systems, the integrative nature of the job suggests that type IV systems should at the very least provide two distinct types of information.

As is illustrated in Fig. 14.2, integrative information systems aim to facilitate the overall co-ordination of a project or programme. They place particular emphasis on the way in which the activities of separate management units jointly contribute to the desired outcome. PERT techniques (project evaluation review techniques) are of this type, as are contract, project and programme planning procedures (Cleland and King, 1975). Coupling information systems, on the other hand, aim to facilitate the co-ordination of particular management units. Whilst of necessity operating within the broader content established by the integrative planning systems, their orientation, scope and hence design can be very different. Many stock control procedures and interdepartmental scheduling procedures are of this type, as are information systems which help to co-ordinate the activities of such separate functions as production and marketing.

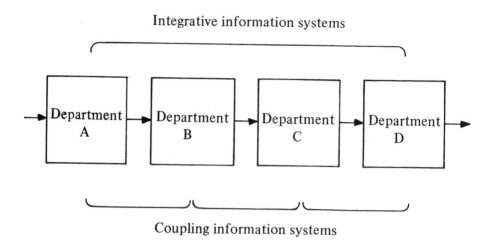

Fig. 14.2   Integrative and coupling information systems

Functionally organised enterprises tend to emphasise the need for type I information systems. There is less overt need for the type II and IV project systems, although increasingly it is being recognised that these can play a valuable role, and even functionally oriented information systems for decision making of the type III variety are often weakly developed at the official level. But what are the information system needs of the matrix organisation? What combination of systems is required to support the operation of a management structure that explicitly aims to integrate the often disparate activities of separate management units?

Information systems for the matrix organisation

It is tempting to think that there might be a single design strategy that would help to identify the information needs of a matrix organisation. Given the emphasis that has been placed on formulating the outline of such a strategy for the functionally managed organisation, surely something similar must be necessary for complementing a matrix structure? However such thinking is misguided. It is increasingly being realised that many of the norms underlying the systematic design of such information systems may not correspond to the realities of organisational life. Several recent investigations have found that high performing enterprises have complex, multi-faceted and even mutually conflicting information systems (Gordon and Miller, 1976; Grinyer and Norburn, 1975). Whilst a concern for simple information system design strategies might meet the designers' needs for clarity, order and tidiness, it is not so obviously related to ensuring organisational effectiveness.

The matrix organisation is complex and multi-faceted. The effectiveness of the whole organisation depends on achieving a continuous balance between vertical and lateral processes. It is this aspect of the matrix approach that provides the clue to its information needs. Rather than looking for a single integrated and tidy information system, it needs to be recognised that to survive and flourish the matrix requires a diversity of information systems. Just as there is a need for both decision and control information, so both the vertical and lateral processes within the matrix need to be provided with their information support. A matrix structure requires, in other words, types I, II, III and IV information systems.

Ways in which type I and II systems may be combined in an integrated control system for matrix organisation have been suggested by Vancil (1973) and Davis (1973), although little evidence of the actual effect of such strategies exists at present. One possibility is that of treating functional departments as cost centres, products or projects as profit centres, with separate controls on each. A second step is to use transfer pricing to allocate part of the profit to the functional managers and part to projects or products (though the frequent arbitrariness of price determination may engender its own forms of conflict). The third possibility is a system of double counting, with total costs or profits being calculated both vertically and horizontally in an accounting matrix which matches the organisational one (cf. Wilkinson, 1974). What evidence we have on the information and control systems that have been associated with the implementation of matrix organisations, and admittedly this is rather limited at present, tends to support the need for a multiple information system strategy.

In the National Aeronautics and Space Administration, surely one of the most complex of modern organisations, project and functional control systems functioned alongside one another. As Sayles and Chandler (1971, p. 307) comment:

> In the daily course of events, managers have to use, and be responsive to the use by others of, a variety of control measures. Some are

consistent with, and critical to, . . . 'dispersed' responsibilities [where the emphasis is on an entire project, without regard to inter-personal, departmental or organisational boundaries]; others represent the more traditional concern with 'jurisdictional' responsibility (of a functional nature) . . . . In the real world of advanced technology implementation both types of controls are necessary.

However, although important, control information alone was not sufficient for the management of this complex enterprise.

Unfortunately, most high- and low-level controls try to measure accomplishment: how actual costs, schedules, and performances compare with the original estimates. This can be misleading, in part because such controls assume rigid plans, compartmentalized jurisdictions and responsibilities. ('Our engineers are mesmerized by watching dollars when they should be watching technology and people.')

The adaptability and responsiveness of the whole organisation depended on the use of more flexible sources of decision oriented information. Given the dominance of the overall mission of this agency, the managers themselves responded to this problem and provided their own sources of information (Sayles and Chandler, p. 314):

Rather than emphasising regular reports, managers place importance on random organisational probes designed to identify co-ordination problems. Thus, each . ... manager concentrates on monitoring the interfaces . . . . These include suppliers of components and subsystems, the next and preceding work-flow stage, test and control groups.

The initial self-designed information-for-decision systems were of an informal nature. With time, however, such pressing needs for information for decision making resulted in the design and official implementation of coupling information systems of the type IV variety, type IV integrative systems which 'direct the attention of . . . managers to the total system's requirements and to the breakdowns that threaten its performance', and, within the functional management units, supportive information systems of the type III variety.

The descriptions of matrix oriented information systems in other governmental agencies and departments also stress the complementary role of project and functional control systems. The very terminology of programme planning and budgeting systems serves to emphasise the recognised need for, if not the actual attainment of, project information systems which serve both decision making (planning) and control (budgeting) needs.

Within the context of the industrial research laboratory, Wilkinson (1974) outlines how functionally oriented financial control systems can be adapted to provide project and programme control information and also considers, albeit briefly, the need for type IV planning information within the matrix structure. Costs were planned and monitored on the basis of individual research projects and project groupings as well as in relation to overall cost type and functional location. Bergen (1975), on the other hand, discusses the construction and use of a type IV project planning system for new products which identifies resource and time requirements by project, function and specific managerial activity. Complementing the existing functional controls, this planning procedure also had the potential of providing the basis for a subsequent type II project control system through which actual accomplishments could be compared with the original plan. In a comprehensive, even if not detailed, description of matrix support systems, Goggin (1974) outlines the multi-faceted management systems that are used to support Dow Corning's 'multi-dimensional' structure. MbO, profit reporting and personnel reviews provide the basis for both type I and II control systems whilst corporate planning, economic evaluation and a comprehensive new business planning system constitute systems of the type III and IV decision varieties.

These and the few other descriptions of practice in the area do provide an indication that a variety of information systems are in simultaneous use in matrix organisations. However, with the possible exception of the more thorough review of the National Aeronautics and Space Administration's experiences provided by Sayles and Chandler (1971), they give little indication of either the way in which the systems actually function as against how they are designed to function, or the relative or combined effectiveness of the various approaches. Yet in this area, as in functional organisations, actual practice can, and does, deviate substantially from what is both intended and claimed.

Problems and potential

Information systems in a matrix organisation should provide flows of information that support the conflicting yet ultimately interrelated pressures that constitute the matrix itself. Yet the very fact that the overall integration of the enterprise is achieved through the use of potentially conflicting means itself provides the basis for some common and important problems.

The different types of information system have different sources of organisational support and different patterns of growth and development. The functional control systems draw their strength from the vertical power hierarchy which they are designed to complement, whilst the use and final impact of the project oriented information systems and particularly the project decision systems depends on the often more fragile patterns of lateral relationships.

To be effective, project control systems need to be associated with an organisational structure and culture that stresses the overriding importance of the goals of the enterprise. Project controls without a project oriented management structure have little effect and whatever may be a particular enterprise's equivalent to 'getting a man on the moon' must be known, visible and well articulated. In order to survive, such systems must be given their own separate existence. Whilst they can, and often should, share a common data base with the functional control systems, project control systems need to be much more than minor extensions of the functional controls. Product costs appended to functional reports or highly summarised project control information alongside detailed functional control information are not sufficient. To be effective, the project control information needs to be at least as visible and well articulated as the functional control information.

If these conditions are fulfilled, however, the use of the type II project control systems is capable of undermining the very different and equally valuable role that can be played by the type IV systems for project decision making. For the planning activities that the type IV systems aim to facilitate can easily be dominated by the anticipation of subsequent evaluation and control on the basis of the type II systems. Where a great deal of emphasis is placed on monitoring actual project performance against the original plans, the plans themselves may be formulated so as to provide easier targets rather than realistic guides to action. Rather than helping to identify potential problem areas, the project planning systems can become mere adjuncts of the control procedures. As the managers seek to establish plans that improve their own performance-potential in an interdependent organisation, the operation of the system can help to create rather than overcome inter-unit conflict.

What is needed is that the decision oriented information systems should become separate from and independent of evaluation oriented control systems, providing local information resources at the operating level (Hedberg, 1975). An increasing number of people are beginning to realise that the involvement of managers in the design of their own decision oriented information systems may be one way of attempting to achieve this. In this way, and particularly if priority is given to the establishment of the decision oriented information systems, both type III and type IV systems might be given a source of independent support. Rather than using the information of 'others' in decision making, managers might be able to establish a sense of ownership over the information that they use. This might provide not only a means of ensuring the viability of these types of information system, but also help to ensure their more effective utilisation.

Ultimately, of course, the final effectiveness of any type of information system or combination of information systems depends on its appropriate use (Hopwood, 1973). It is easy for complex information systems to be either rejected or taken at their face value. But, to have the appropriate effect, they must be used in a careful and considered manner, sometimes being used but on other occasions being set aside in the light of alternative evidence. How information is used reflects quite general managerial characteristics. To change these is difficult. Certainly educational programmes have a role to play if they aim to help managers understand the rationale, advantages *and* disadvantages of the systems rather than merely present a public relations version of the activities of the information specialists. But real progress might depend on trying to influence how managers actually use information. As information system designers try to do this, their efforts will be related directly to the organisational development activities of those organisational designers who strive to create the social environment in which a matrix structure can survive and flourish.

Conclusion

The design of information systems for matrix organisations remains an emerging area of concern. Many unknowns and doubts remain and, as yet, many of the early experiments have still to endure the test of practice. At this stage, however, there is reason to think that the appropriate systems will reflect the conflicting but mutually reinforcing pressures of the matrix itself. Rather than constraining the simultaneous exercising of vertical and horizontal influences, such information systems will provide both the means for these influences to function and some of the checks and balances that are so necessary for the enterprise as a whole to function effectively. However alien design principles of this type may be to those information system designers who look for orderliness and simplicity, they do offer the possibility of nurturing and supporting the vital organisational processes that the matrix approach seeks to promote.

Summary

The effectiveness of matrix organisation is often frustrated by the retention of inappropriate information and control systems based on a previous functional structure. An example is described in which a reorganisation aimed at strengthening lateral co-ordinating links was made ineffective by a new information system which placed greater emphasis on the vertical reporting patterns of separate departments.

To assist in the development of information systems which will support rather than hinder the operation of matrix structures a two-dimensional classification of information systems is proposed, which distinguishes between control and decision oriented information and between functional emphasis on resource utilisation and project emphasis on end results. This results in four types of information system being distinguished. An examination of the needs of matrix organisations suggests that they require the simultaneous existence of all four types of system, rather than a single overriding design strategy. The little evidence which is available tends to support this view.

The danger that decision oriented project information systems may be distorted by being subordinated to evaluation oriented control systems may be avoidable by involving managers in the design of their own decision information systems. This might make it more likely that information systems will be used constructively to increase the effectiveness of matrix organisations.

15   Multi-dimensional management

Urgent questions

If our small survey is any guide then more organisations are at present contemplating or in process of introducing matrix structures than are thinking of abandoning them. If this is the case then it is becoming a matter of some urgency to find better answers than are available so far to a number of the questions raised in this book.

Effectiveness

We do not really know how well, in the majority of cases, matrix structures achieve the various objectives for which they are designed or what are the conditions of success. It may be that this knowledge is not essential. A conservative rule of thumb would be – do not embark on matrix management unless you have to. But some reasons for introducing matrix structures are forward-looking ones, rather than responses to present crisis; the objective which we described as adaptation is a case in point. This kind of decision really ought to be made with better knowledge of the prospects of success.

Operating problems

The idea of 'problems' is one of the basic categories in which we think about management. Of course there are problems, life is full of them. Ask any manager – or indeed any human being – what problems he has, and he will recite an impressive list. So the reports of problems in matrix are not a surprise. The question is how real these problems are. By how much do they exceed the normal everyday problems of any management structure? Which of these problems are the really painful ones, and under what circumstances do they arise? And most important of all: can they be solved, or lived with, and, if so, how?

Implementation

At the moment, the approach to implementing matrix organisations does not seem to be guided by any explicit theory or understanding of the factors involved, but seems rather to depend on the style and preferences of the managers taking the decision. Some are unilateralists by nature and inclination, others bilateralists. Does the method of implementation really matter to the success of matrix? It would be useful to have some evidence.

Definition of roles

The argument about ambiguity and flexibility is by no means settled. Statements both in support and in condemnation of careful definition of authority and accountability relationships are made with great assurance, each backed by a whole philosophy. But do these definitions, or the lack of them, really help or hinder the operation of a matrix? Or does it not matter?

People

Does matrix operation call for a special breed of managers and employees? What personal qualities and capacities are needed by people who take up these roles? How can they be identified and developed?

A practical approach

Decisions about organisation have to be made under the pressure of events; they cannot wait until all the evidence has been assembled – if it ever is. It appears that the pressures that have led, and are leading, organisations to consider multi-dimensional structures are on the increase, and, as we emerge from our exploration of matrix management, we should be able to suggest what practical approach managers can adopt right now. The following seem to be essential steps for a management group or individual facing the question of whether to introduce or formalise a matrix structure.

1 *Decide what aims adopting the structure is intended to achieve* Better co-ordination? Improved use of resources? Clear accountability for project or product objectives? Increased expertise? Better opportunities for staff? Ability to adjust to external changes? Or a mixture of several aims? Once this is clear, it becomes possible to identify the particular characteristics the matrix structure needs to have in order to do the job it is intended for.

2 *Assess the present character of the organisation and its members* Is it dynamic, flexible, easy to change? Are people committed to getting results? Do they attach importance to clear roles and responsibilities,

rules and procedures? Are managers (starting at the top) reluctant to delegate? How do people react to threats to their power? If the change involves setting up new roles, who has the ability to take them on? By making an assessment of the present organisational culture and its human capital it becomes easier to see what types of structure have a large or little chance of success, or to plan a phased transition to the desired arrangement.

3 *Decide on the most appropriate form of matrix management* Assuming that these initial steps have not produced second thoughts about the intended change, the next question is just what form of multi-dimensional structure is going to be most appropriate to achieve the objectives that have been defined, given the current organisation, its culture and the people in it? Some of the options available, and the particular ways in which they relate to the main decision factors, have been discussed earlier.

4 *Consider the likely impact of the chosen option on the responsibilities and power of the people they will affect* Will the people whose responsibilities are to be changed be able to cope? Which individuals, which groups, are likely to gain, which to lose out? How can they be expected to react? What kind of conflicts is the structure going to set up? Unless these questions are carefully considered at the outset, and a strategy is devised for dealing with them, serious disruption may occur.

5 *Decide how the change should be implemented* Is a unilateral decision by top management the best way? Will it produce the desired commitment to the new structure? Or should a decision be taken jointly with the managers affected? Should non-managerial staff be involved? A decision in favour of a bilateral or participative implementation will, of course, imply going over the earlier ground again, in order to arrive jointly at a preferred structure.

6 *Take a conscious decision whether and how to define role relationships in the new structure* This is a key issue which should not go by default. Should accountability and authority relationships be clearly spelt out, or left open? What processes should be adopted for this purpose? It is also important that the decision on this should be understood and accepted by the people whose roles are at issue.

7 *Assess training and development needs* What new knowledge, awareness and skills are needed to operate successfully in the new structure? Do team-working skills, influencing skills, conflict resolution skills need to be developed? Whether it is decided that these issues can be handled by internal team building programmes, or that outside

training facilities need to be used, appropriate resources have to be identified and deployed.

8 *Examine compatibility of management systems* Are existing information and control systems geared to the new structure? Do they support or discourage the types of behaviour required? What new information systems are needed? Similarly, selection methods and reward systems, including appraisal, management development, promotion, job evaluation and pay, need to be reviewed and made compatible with the matrix structure.

9 *Review the operation of the structure and modify it where necessary* This is an obvious requirement. Only practice will show whether the assumptions on which the change was based are justified by results.

It is suggested that these are essential steps in introducing or reviewing a matrix organisation. Obviously they pose many difficult questions, calling for the exercise of discretion and judgement. This book has attempted to assemble relevant arguments and information on which answers can be based.

Tomorrow's organisation?

For some of its more enthusiastic advocates, matrix is the organisation of the future, and the fact that more firms seem to be adopting it is part of a welcome and necessary trend. Other managers, more sceptical in their outlook, look with a jaundiced eye at what they see as merely another in a long series of fads and fashions sweeping across the management scene at ever decreasing intervals. Some of these fads have been relatively harmless – a way of keeping a few specialists in employment without having any noticeable impact on where the work gets done. But matrix management is not like that; as we have seen it can create real headaches for managers who feel they have enough to worry about as it is. The question of whether matrix management responds to any genuine new social and organisational needs is therefore one of some importance. For the particular organisation, it is linked with the even more difficult question of deciding whether these needs, if they are genuine, apply to *us*. In simple terms – is matrix management necessary?

Conflicting trends

Before deciding that this question is impossible to answer it is worth looking briefly at two contrasting trends in the economic and political

212

sphere, in society and in management which have provided the context, and perhaps the underlying rationale, for the emergence of matrix organisation. This contrast can perhaps be shown most easily by drawing attention to the increasing and dramatic divergence between the classical economic forces of supply and demand.

On the supply side it is obvious that in advanced industrial societies the provision of goods and services by both industry and the state is being handled by increasingly large units, using technologies which call for massive investment, and attempting, with a great deal of difficulty, to integrate the work of large numbers of people. The scale of these undertakings, and particularly of the capital base from which they operate, leads to attempts to centralise control, to use large, comprehensive administrative and accounting systems and to process and integrate information through bigger, faster computers and communication networks. This is not only the case in business, with the emergence of ever larger concentrations of capital and people, both within and across national boundaries. The reorganisation of local government and the health service and the amalgamation of steel and of airways provide parallel recent examples in the United Kingdom public sector of this drive to create very large administrative units to supply social and consumer needs.

On the demand side, however, the movement has been very much in the opposite direction. The pressure for 'market orientation' in business is a reflection, by now familiar, of the trend towards diversification of consumer preferences. The market stance of 'any colour you like as long as it's black' has long been as obsolete as the Model T. Similar trends are gathering momentum in politics and society, where the pressures are for local autonomy and devolution of power, for employee and community participation in decision making, for sensitivity to the needs of disadvantaged and minority groups and to the demands of environmental lobbies, conservationist interests, consumer organisations and a host of others. The overall picture is one of concentration of supply confronting diversification of demand – the immovable object and the irresistible force. Something has to give: could it be organisational simplicity?

Similar contrasts can be seen in the development of management thought and practice. Specialisation and division of labour within the departmentalised functional organisation has been the instrument for increased productivity and the build-up of large concentrations of labour and capital. It has tended to be accompanied by more routine work, simplification, restriction and standardisation of working relationships and increasing rigidity of organisational structures. The idea, now gaining ground, that these trends are in the process of being reversed may be an

213

illusion: many signs point the other way, but it is too early to tell. However, there is no doubt that a powerful opposing trend is making itself felt. Whatever managerial practice may be, the conventional wisdom is veering towards stressing the importance of the face-to-face working group, the value of autonomy, job enrichment, participative management, 'Theory Y', quality of working life and the rest.

No doubt the diversification of demand and the greater sophistication of technology are also playing their part in this. The extreme fragmentation and routinisation of work associated with 'scientific management' yields high productivity in the mass-manufacture of uniform products like the proverbial widget. The proportion of industrial work which this now represents must be shrinking year by year. Consumer choice, market orientation and the premium on novelty lead to short product life cycles in consumer industries, while technological sophistication and change place emphasis on development rather than long run production for industrial markets. For increasing sectors of industry the 'project' model of organisation is replacing the 'production' one. Even this is no doubt an oversimplification. The actual situation presents a continuum from the ongoing to the unique – from widgets to cyclotron. The question we are faced with is whether we stand to gain more, in any particular context, from a production or a project definition of tasks, and if we need both, whether unidimensional organisation can cope?

Perhaps this can help to find answers to the question whether matrix management is necessary. Obviously there is no single overriding answer for all organisations. The two conflicting trends towards concentration of supply and diversification of demand are reflected by two ways of ordering resources in the organisation. Efficient supply favours structuring along the dimension of process and specialisation, effective response to demand puts a premium on purpose and project organisation. As long as a structure based on one of these dimensions can satisfy both sets of requirements there is no need to look any further. It is when the two needs can no longer be reconciled within a unitary structure that some form of matrix management apparently becomes unavoidable. Indeed, Timmermann (1971) describes the function of a matrix in precisely these terms: it is an attempt, he says, to incorporate the market mechanism of supply and demand into the structure of the organisation.

Having arrived at such an 'answer', all we have done, of course, is to raise a new question. If a unidimensional structuring of responsibilities cannot reconcile supply and demand, what can? Timmermann's statement can be taken to imply that matrix organisation has a single meaning, whereas we have, by now, seen a wide enough range of options to be convinced that

214

this is not so. Our initial decision not to commit ourselves to a strict definition of matrix organisation was taken so as not to restrict exploration. Having explored the field it seems to me that any but a very broad definition would not be helpful. Matrix management means an *approach* to management which institutionalises dual or multiple influence, including a measure of divided authority. In most cases it involves the use of a matrix organisation. Matrix organisation, in turn, is a *class* of organisation which supplements multiple authority by multiple membership. Beyond this it is a question of choosing and putting into effect the most appropriate pattern of relationships to meet the constraints of supply and demand in ways that are compatible with the capacities of the people involved.

Multiple membership and human flexibility

Matrix management may be necessary in some circumstances, but is it possible? If the pressures under which they operate create for some organisations a need for multi-dimensionality in the way they structure interactions, communications and patterns of accountability, can the people in these organisations rise to such demands or are the expectations inherent in the matrix concept too unrealistic? Can people live with multiple group membership and divided responsibility?

The information we have is contradictory and inconclusive. Most of us operate very happily as members of a series of different groups – family, club, working group, neighbourhood and so forth. We switch with great ease from one role to another. But the matrix situation is different when it involves reconciling membership of different groups within the same role, and the research on organisational stress (Kahn et al., 1964) suggests that it is these 'boundary' roles which impose particular strains. Such roles, stressful or not, are of course common in every organisation; perhaps we have simply learnt to cope better with the long-established boundary roles, such as salesmen, foremen or specialist advisers, than with the newer ones created by matrix organisations.

It may be also that flexibility varies not only between persons, but also between situations. Psychological research suggests that behaviour under conditions of threat and crisis is very much more limited and stereotyped than in a supportive and secure environment. The same people who are capable of risk-taking and versatility in a relaxed and stimulating environment may become rigid and unco-operative in an atmosphere of restriction and fault-finding.

Perhaps this is an area where prophecies are more self-fulfilling than in most. Management philosophies which emphasise human limitations – 'scientific management' is the prime example – seem to have led to the growth of many large organisations which to the outsider seem dauntingly cumbersome, inflexible and muscle-bound and in which it is exhaustingly troublesome to get the simplest things done if they run counter to the inertia of established practices. Looking at such organisations one asks oneself – if this is how ordinary bureaucracies behave, what hope is there for matrix?

But, at the same time as it has witnessed a frightening growth in organisational incompetence, this century has also seen some astonishing organisational feats. One thinks, for instance, of the vast co-ordination of activities that made possible the Normandy invasion or the first landing on the moon. Between these extremes one is aware of many organisations which are managing to face up to complexity and multi-dimensionality with varying degrees of success, and for the present we can only speculate why some manage better than others. The range of human response to complex environments is apparently enormous – from effortless adaptation to catatonia.

But these are not random outcomes. Whether the matrix structures of the future prove to be unmanageable, accident-prone and paralysed by conflict, or whether they are effective in achieving work objectives and creative in adapting to change will depend, in my opinion, on the care with which the options are analysed and the success with which they are matched to the needs of the situation as well as to the capacities and potential of the people working in them.

Summary

Although more information about the operation and effectiveness of matrix management is urgently required, it is possible to set down a series of practical steps for those contemplating a matrix approach. They include:

    deciding aims;
    assessing present organisation;
    selecting form of matrix;
    predicting effect on individuals;
    deciding how to implement;
    deciding whether or not to define roles;

training and development;

examining management systems;

reviewing.

Before embarking on such an approach, it is important to decide whether matrix management corresponds to real needs or merely represents a passing fashion. Two conflicting trends, towards large scale concentration of resources on the one hand, and increasing diversification of demand on the other, are apparent in the environment. Multi-dimensional structuring of organisations seems to become a necessity when a unitary structure can no longer cope with the conflicting requirements created by these trends.

Whether human beings can adjust to the demands made by such organisational complexity remains an open question, but the assumptions made by different management philosophies may turn out to be self-fulfilling, and a careful analysis of options and capacities is essential.

Bibliography

Aiken, M. and Hage, J., 'The Organic Organisation and Innovation', *Sociology*, vol. 4, 1970.

Algie, J., 'Management and Organisation in the Social Services', *British Hospital Journal and Social Services Review*, vol. LXXX, no. 4184, 26 June 1970, pp. 1245-8.

Anthony, R. N. and Herzlinger, R., *Management Control in Non Profit Organisations*, Irwin, 1975.

Argyris, C., 'Today's Problems with Tomorrow's Organisations', *Journal of Management Studies*, vol. 4, no. 1, February 1967, pp. 32-55.

Argyris, C., *The Applicability of Organisational Sociology*, Cambridge University Press, 1972.

Bennis, W. G., 'Changing Organisations', *Journal of Applied Behavioural Science*, vol. 2, no. 3, 1966, pp. 247-63.

Bergen, S. A., 'The New Product Matrix', *R and D Management*, vol. 5, no. 2, 1975.

Bernhard, A., 'Attraktive Möglichkeiten der Matrix-Organisation' (Attractive possibilities of matrix organisation), *Industrielle Organisation*, vol. 43, no. 8, 1974.

Bion, W. R., *Experiences in Groups and Other Papers*, Tavistock, London, 1961.

Brings, K., 'Erfahrungen mit der Matrixorganisation' (Experiences with matrix organisation), *Zeitschrift für Organisation*, 2/1976.

Brooke, M. Z. and Remmers, L., *The Strategy of Multinational Enterprise*, Longman, London, 1970.

Brooks, P. W., 'Management and Marketing in Large Enterprises', *Aeronautical Journal*, vol. 74, no. 720, 1970, pp. 936-47.

Brown, W., *Exploration in Management*, Heinemann Educational Books Ltd, London, 1960.

Brown, W., *Organization*, Heinemann Educational Books Ltd, London, 1971.

Brown, W. and Jaques, E., *Glacier Project Papers*, Heinemann Educational Books Ltd, London, 1965.

Brunel Institute of Organisation and Social Studies (BIOSS), Health and Social Services Organisation Research Units, 'Collaboration Between Health and Social Services', Working Paper, 1976.

Burns, T., 'Models, Images and Myths', in Gruber, W. H. and Marquis, D. G., *Factors in the Transfer of Technology*, The MIT Press, Cambridge, Mass., 1969.

Burns, T. and Stalker, G. M., *The Management of Innovation*, Tavistock, London, 1961.

Child, J., 'Managerial and Organisational Factors Associated with Company Performance', *Journal of Management Studies*, vol. 11, 1974, pp. 175–89, and vol. 12, 1975, pp. 12–27.

Cleland, D. I., 'The Deliberate Conflict', *Business Horizons*, February 1968, pp. 78–80.

Cleland, D. I. and King, W. R., *Systems Analysis and Project Management*, McGraw-Hill, New York, 1968, Second edition 1975.

Cleland, D. I. and King, W. R., *Systems, Organisations, Analysis, Management : A Book of Readings*, McGraw-Hill, New York, 1969.

Cleland, D. I. and King, W. R., 'Organising for Long Range Planning', *Business Horizons*, August 1974, pp. 25–32.

Corey, E. R. and Star, S. H., *Organisation Strategy : A Marketing Approach*, Division of Research, Graduate School of Business Administration, Harvard University, Boston, Mass., 1971.

Crozier, M., *The Bureaucratic Phenomenon*, Tavistock, London, 1964.

Dalton, G. W., Barnes, L. B. and Zaleznik, A., *The Distribution of Authority in Formal Organisations*, The MIT Press, Cambridge, Mass., 1968.

Davis, S. M., 'Matrix Organisation and Behaviour in Domestic and Multi-national Enterprise', Graduate School of Business Administration, Harvard University, April 1973. (Also obtainable from Case Clearing House of Great Britain and Ireland.)

Department of Health and Social Security, '*Management Arrangements for the Reorganised National Health Service*', HMSO 1972.

Dow Corning Corporation, *The Multidimensional Organisation*, 1973.

Drucker, P., 'New Templates for Today's Organisations', *Harvard Business Review*, January/February, 1974.

Eccles, A. J., 'Serving Two Masters in the Modern Organisation', *Management Decision*, vol. 12, no. 5, 1974.

Emery, F. E. *Characteristics of Socio-Technical Systems*, Tavistock Institute of Human Relations Document no. 527, 1959. Reprinted (in part) in Davis, L. E. and Taylor, J. C. (eds), *Design of Jobs*, Penguin Books, Harmondsworth, England, 1972.

Fayol, H., *General and Industrial Management*, Pitman, London, 1949. (Translated from the original *Administration Industrielle et Générale*, 1916.)

Frankel, D. S., *A Behavioural Analysis of the Advertising Agency : Behavioural Determinants of Effectiveness in Advertising Agency Account Groups*, unpublished PhD thesis, London Graduate School of Business Studies, 1975.

Friedson, E., *Profession of Medicine*, Dodd, Mead, New York, 1970.

Fulmer, R. M., 'Product Management: Panacea or Pandora's Box', *California Management Review*, vol. VII, no. 4, Summer 1965. Reprinted in Cleland, D. I. and King, W. R., 1969, *op.cit.*

Galbraith, J. R., 'Matrix Organisation Designs', *Business Horizons*, February 1971, pp. 29–40.

Galbraith, J. R., *Designing Complex Organisations*, Addison-Wesley, Reading, Mass., 1973.

Getzels, J. W and Jackson, P. W., *Creativity and Intelligence*, John Wiley, London, 1962.

Goggin, W. C., 'How the Multidimensional Structure Works at Dow Corning', *Harvard Business Review*, January/February, 1974.

Goldstein, H., *Social Work Practice – A Unitary Approach*, University of Carolina Press, Columbia, Carolina, 1973.

Goodman, R. A., 'Ambiguous Authority Definition in Project Management', *Academy of Management Journal*, vol. 10, 1967, pp. 395–407.

Gordon, L. A. and Miller, D., 'A Contingency Framework for the Design of Accounting Information Systems', *Accounting, Organisations and Society*, vol. 1, no. 1, June 1976, pp. 59–70.

Gray, J. L. 'Matrix Organisational Design as a Vehicle for Effective Delivery of Public Health Care and Social Services', *Management International Review*, vol. 14, part 6, 1974, pp. 73–87.

Greenwood, R. and Stewart, J. D., 'Corporate Planning and Management Organisation', *Local Government Studies*, October 1972, pp. 24–40.

Greenwood, R. and Stewart, J. D., *Corporate Planning in English Local Government*, Charles Knight & Co. Ltd, London, 1974.

Greiner, L. E., 'Evolution and Revolution as Organisations Grow', *Harvard Business Review*, July/August, 1972.

Grinyer, P. H. and Norburn, D., 'Planning for Existing Markets : Perceptions of Executives and Financial Performance', *Journal of the Royal Statistical Society*, Series A, vol. 138, Part I, 1975, pp. 70–97.

Gulick, L. and Urwick, L. (eds), *Papers on the Science of Administration*, Columbia University, Institute of Public Administration, New York, 1937.

Handy, C. B., *Understanding Organisations*, Penguin Books, Harmondsworth, England, 1976.

220

Harvey, E., 'Technology and the Structure of Organisations', *American Sociological Review*, vol. 33, 1968.

Health Services Organisation Research Unit, 'Working Papers on the Reorganisation of the National Health Service – Revised October 1973', Brunel University, 1973.

Hedberg, B., 'Computer Systems to Support Industrial Democracy' in Mumford, E. and Sackman, H. (eds), *Human Choice and Computers*, North-Holland, 1975, pp. 211–30.

Hendry, W. D., 'A General Guide to Matrix Management', *Personnel Review*, vol. 4, no. 2, Spring 1975, pp. 33–9.

Herzberg, F., *Work and the Nature of Man*, Harcourts, Brace and World, 1966.

Hill, R., 'Corning Glass Reshapes its International Operations', *International Management*, October, 1974.

Hinings, C. R., Hickson, D. J., Pennings, J. M. and Schneck, R. E., 'Structural Conditions of Intraorganizational Power', *Administrative Science Quarterly*, vol. 19, no. 1, March, 1974.

Hobbs, C., 'The Benefits and Problems of R and D in Contracting', *Construction R and D Journal*, no. 3, 1969.

Home Office et al., *Report of the Committee on Local Authority and Allied Personal Social Services*, (Seebohm – Chairman) Cmd. 3703, HMSO, London, 1968.

Hopkins, D. S., *Options in New-Product Organisation*, Report No. 613, The Conference Board, New York, 1974.

Hopwood, A. G., *An Accounting System and Managerial Behaviour*, Saxon House, Farnborough, England, 1973.

Hudson, L., 'Personality and Scientific Aptitude', *Nature*, vol. 198, 1963, pp. 913–14.

Janger, A. R., 'Anatomy of the Project Organisation', *Business Management Record* (National Industrial Conference Board), November, 1963.

Jaques, E., *A General Theory of Bureaucracy*, Heinemann Educational Books Ltd, London, 1976.

Jay, A., *Management and Machiavelli*, Hodder and Stoughton, London, 1967.

Kahn, R. L., Wolfe, D. M., Quinn, R. P., Snoek, J. D. and Rosenthal, R. A., *Organisational Stress : Studies in Conflict and Ambiguity*, John Wiley, New York, 1964.

Kast, F. E. and Rosenzweig, J. E., *Organisation and Management : A Systems Approach*, McGraw-Hill, New York, 1970.

Kingdon, D. R., *Matrix Organisation*, Tavistock, London, 1973.

Knight, K., 'Matrix Organisation : A Review', *Journal of Management Studies*, vol. 13, no. 2, May 1976.

Knight, K., 'Authority and Responsibility in the Matrix Organisation', *R and D Management*, vol. 7, no. 3, June, 1977.

Lawrence, P. R. and Lorsch, J. W., *Organisation and Environment*, Division of Research, Graduate School of Business Administration, Harvard University, Boston, Mass., 1967.

Lawrence, P. R. and Lorsch, J. W., 'New Management Job : the Integrator', *Harvard Business Review*, November/December, 1967 (a), pp. 142-51.

Likert, R., *New Patterns of Management*, McGraw-Hill, New York, 1961.

Likert, R., *The Human Organisation*, McGraw-Hill, New York, 1967.

Lorsch, J. W. and Lawrence, P. R., *Organisation Planning : Cases and Concepts*, Irwin-Dorsey, Homewood, Ill., 1972.

Ludwig, S., 'Should Any Man Have Two Bosses?', *International Management*, April 1970.

Mann, J., 'Matrix Management', *Local Government Studies*, June 1973, pp. 31-7.

Marquis, D., 'Ways of Organising Projects', *Innovation*, no. 5, 1969, pp. 26-33.

Massie, J. L., 'Management Theory' in March, J. G. (ed.), *Handbook of Organisations*, Rand McNally, Chicago, 1965.

McKelvey, B., 'Guidelines for the Empirical Classification of Organisations', *Administrative Science Quarterly*, vol. 29, no. 4, December 1975, pp. 509-25.

Melcher, R. D., 'Roles and Relationships : Clarifying the Manager's Job', *Personnel*, vol. 44, no. 3, May/June, 1967, reprinted in Cleland, D. I. and King, W. R., 1969, op.cit.

Miller, E. J. and Rice, A. K., *Systems of Organisation*, Tavistock, London, 1967.

Milner, C. G., 'Innovation in Shipbuilding', *R and D Management*, vol. 2, no. 1, 1972.

'The New Local Authorities : Management and Structures' (often referred to as the 'Bains Report'), HMSO, 1972.

Paskins, D. C., *'Dynamic Group Responses and the Effects upon a Matrix Organisational Structure in a Contract Industry'*, unpublished MPhil thesis, Brunel University, 1977.

Perham, J., 'Matrix Management : A Tough Game to Play', *Dun's Review*, August 1970, pp. 31-4.

Peterson, A. W., 'Planning, Programming and Budgeting in the G.L.C. – What and Why', *Public Administration*, vol. 50, Summer 1972, pp. 119-26.

Pettigrew, A. M., 'Information Control as a Power Resource', *Sociology*, vol. 6, no. 2, 1972, pp. 187–204.

Pettigrew, A. M., *The Politics of Organisational Decision-making*, Tavistock, London, 1973.

Pincus, A. and Minahan, A., *Social Work Practice – Model and Method*, F. E. Peacock, Illinois, 1973.

Porter, L. W., Lawler, E. E. and Hackman, J. R., *Behaviour in Organisations*, McGraw-Hill, New York, 1975.

Pugh, D. S., Hickson, D. J. and Hinings, C. R., *Writers on Organisations*, Second Edition, Penguin Books, Harmondsworth, England, 1971.

Pugh, D. S. and Hickson, D. J., *Organizational Structure in its Context: The Aston Programme I*, Saxon House, Farnborough England, 1976.

Pugh, D. S. and Hinings, C. R., *Organizational Structure: Extensions and Replications: The Aston Programme II*, Saxon House, Farnborough England, 1976.

Pugh, D. S. and Payne, R., *Organizational Behaviour in its Context: The Aston Programme III*, Saxon House, Farnborough England, 1977.

Rice, A. K., *Productivity and Social Organisation: the Ahmedabad Experiment*, Tavistock, London, 1958.

Roeber, J., *Social Change at Work: The ICI Weekly Staff Agreement*, Duckworth, London, 1975.

Roethlisberger, F. J. and Dickson, W., *Management and the Worker*, Harvard University Press, Cambridge, Mass., 1939.

Rowbottom, R. W., *Social Analysis*, Heinemann Educational Books Ltd, London, forthcoming.

Rowbottom, R. W. et al., *Hospital Organisation*, Heinemann Educational Books Ltd, London, 1973.

Rowbottom, R. W. and Billis, D., 'The Stratification of Work and Organisational Design', *Human Relations*, vol. 30, no. 1, 1977.

Rubin, I. M. and Seelig, W., 'Experience as a Factor in the Selection and Performance of Project Managers', *IEEE Transactions on Engineering Management*, September 1967, pp. 131–5.

Rush, H. M. F., 'The Systems Group of TRW, Inc.' in *Behavioural Science*, National Industrial Conference Board, Personnel Policy Study, no. 216, 1969, pp. 157–71.

Sanders, G., 'Some Thoughts on Project Groups', *Industrial and Commercial Training*, June 1976.

Sayles, L. R. and Chandler, M. R., *Managing Large Systems*, Harper and Row, New York, 1971.

Schemkes, H., 'Kompetenzabgrenzung bei der Mehrlinienorganisation'

(Defining responsibilities in the multi-line organisation), *Zeitschrift für Organisation*, no. 8, 1974.

Seglow, P., 'Conflict and Compromise', unpublished paper, Brunel University Management Programme, 1974.

Shull, F. A., Delbecq, A. L. and Cummings, L. L., *Organisational Decision Making*, McGraw-Hill, New York, 1970.

Social Services Organisation Research Unit (SSORU), Brunel University, *Social Services Departments : Developing Patterns of Work and Organisation*, Heinemann Educational Books Ltd, London, 1974.

Stamp, G., *A Study in Matrix Management*, unpublished dissertation, Brunel University, 1974.

Steiner, G. A. and Ryan, W. G., *Industrial Project Management*, Macmillan, New York, 1968.

Stopford, J. M. and Wells, L. T., *Managing the Multinational Enterprise*, Longman, London, 1972.

Thamhain, H. J. and Wilemon, D. L., 'Diagnosing Conflict Determinants in Project Management', *IEEE Transactions on Engineering Management*, vol. EM–22, no. 1, February 1975.

Thom, N., 'Zur Leistungsfähigkeit der Projekt-Matrix-Organisation' (On the performance capabilities of the project-matrix organisation), *Industrielle Organisation*, vol. 42, no. 3, 1973.

Thompson, J. D., *Organisations in Action*, McGraw-Hill, New York, 1967.

Timmermann, M., 'Matrix-Management' (in German), *Industrielle Organisation*, vol. 40, no. 7, 1971.

Toffler, A., *Future Shock*, Bodley Head, 1970.

Trist, E. L. and Bamforth, K. W., 'Some Social and Psychological Consequences of the Longwall Method of Coal Getting', *Human Relations*, vol. 4, no. 1, 1951, pp. 3–38.

Trist, E. L., Higgin, G. W., Murray, H. and Pollock, A. B., *Organisational Choice*, Tavistock, London, 1963.

Vancil, R. F., 'What Kind of Management Control Do You Need?' *Harvard Business Review*, March/April 1973.

Vickers, G., *Making Institutions Work*, Associated Business Programmes, London, 1973.

Videlo, D. A., 'The Engineering Department Matrix Organisation', *R and D Management*, vol. 6, no. 2, 1976, pp. 73–6.

Walker, A. H. and Lorsch, J. W., 'Organisational Choice : Project vs. Function', *Harvard Business Review*, November/December 1968, pp. 129–38.

Wearne, S. H., 'Project and Product Responsibilities in Industry', *Management Decision*, Winter 1970, pp. 32–5.

Wiggins, J. W., 'Operational Flexibility through Project Management', *National Conference Board Record*, vol. 4, 1967, pp. 68–70.

Wilemon, D. L., 'Transferring Space Age Management Technology', *National Conference Board Record*, vol. 7, 1970, pp. 50–5.

Wilemon, D. L., 'Managing Conflict in Temporary Management Systems', *Journal of Management Studies*, vol. 10, no. 4, 1973, pp. 282–96.

Wilemon, D. L. and Cicero, J. P., 'The Project Manager – Anomalies and Ambiguities', *Academy of Management Journal*, vol. 13, September 1970, pp. 269–82.

Wilemon, D. L. and Gemmill, G. R., 'Interpersonal Power in Temporary Management Systems', *Journal of Management Studies*, vol. 8, no. 4, 1971, pp. 315–28.

Wilkinson, J. B., 'Management Structure in an Industrial Research Laboratory', *R and D Management*, vol. 4, no. 3, 1974, pp. 135–9.

Woodward, J., *Industrial Organisation : Theory and Practice*, Oxford University Press, 1965.

Woodward, J. (ed.), *Industrial Organisation : Behaviour and Control*, Oxford University Press, 1970.

Index

174; sapiential 174; sources of
174-5, 176; in team 87; and
uncertainty 175
Problem solving 128
Problems of matrix management 161-9:
administrative 168, 171; approaches
to 170-1; balance 162-4;
individual 166-8; relationship
164-6
Process, process consultation 186, 188,
189
Product (programme) management 1,
16, 39, 111, 146
Productivity 47
Profitability 66
Programme areas 19-20, 121
Programme evaluation and review
technique (PERT) 201
Programme planning (and budgeting)
systems 19, 200, 201, 204
Project: commitment 33, 37; inter-
disciplinary 148; as model of
organisation 214; planning,
programming of 31; termination of
37; see also Team
Project expediter 144-5
Project management 1, 16: in local
government 19; in R & D 17;
as training 35
Project manager (leader): authority and
power of 6, 143, 163, 176; influence
of 35; responsibility–authority gap
143; role of 27-8, table 3.1, 35,
64-5, 66-7, 176; selection of 176,
181
'Project within function' 12 (n.3), 148,
151-2
Pugh, Derek 139 (n.1, 2)

Remmers, L. 22 (n.7)
Relationship(s): dotted line 7, 12 (n.4),
16, 97, fig. 7.2, 150, 172; between
functional and project managers
35-6; influencing, options 173;
lateral (horizontal) 3, 4, 82, 84,
117 (n.2), 127, 196-7, 198; problems
of 162, 164-6; vertical 3, 4, 82, 84;
role 8, 172-4
Research and development (R & D):
characteristics of 24-5; definition of
23; matrix in 15, 17, 22 (n.1), 24,
26-43, 165, 167; organisational
requirements of 25-6; transfer of

ideas from 39
Resource manager 64, 65, 67, 69
Resources: allocation of 25-6, 66;
slack 3; utilisation of 25-6, 46, 51,
93, 113, 114, 199
Responsibility: and authority 87, 88,
171-2, 174; clarifying 87-9;
definition 49, 172-4, 183 (n.1), 88;
levels of 89
Responsibility chart 49, 173, 191,
fig. 13.1
Reward 51, 55
Rice, A. K. 44 (n.1), 139 (n.5, 6)
Roeber, Joe 58 (n.1)
Roethlisberger, F. J. 139 (n.6)
Role(s): boundaries of 96; clarification
87-9, 141, 191; confusion 187;
definition of 3, 56, 163, 172-4, 211;
in bilateral implementation 173, and
discretion 173, and stress 167;
negotiation of 191; separation of
64; vertical, redefined 163
Role ambiguity 167
Role conflict 55, 167
Rosenthal, R. A. 221
Rosenzweig, J. E. 139 (n.1), 221
Rowbottom, Ralph 90 (n.4, 5), 156
(n.1, 2)
Rubin, I. M. 36
Rush, H. M. F. 22 (n.4), 132, 157,
(n.10)
Ryan, W. G. 22 (n.4), 156 (n.3), 183
(n.2)

Sadler, Philip 117 (n.5)
Sanders, G. 156 (n.3)
Satisfaction 80, 132
Sayles, Leonard 22 (n.4), 152, 203-4,
205
Schemkes, H. 22 (n.2), 183 (n.1)
Scicon (Scientific Control Systems Ltd)
59-72, 109, 113, 122, 127, 128, 132,
150, 177
Scientific management 120, 214, 215
Secondment 88, 103, 151, 152
Secondment matrix 142, 150-1;
variants of 151-2
Seebohm report 94
Seelig, W. 36
Seglow, Peter 165
Selection: for matrix organisation 181;
of project managers 176, 181; test,
16PF 183 (n.8)

231

Self contained units 3, 48, 51: degree of self-containment 50
Semi-autonomous work group 84, 133, 144
Service giving 39, 145, 148, 151
Sheane, Derek 109, 128, 133, 162, 166, 173, 176, 179
Size of organisation 48, 51, 195: and matrix working 31; and structure 56, 123
Sloan School 110, 134
Smith, Adam 120
Social analysis 141, 147, 150, 151, 173
Social effectiveness 114, 115, 131–5, 155; of matrix organisation 134–5
Social services departments 122: development of 91–3; functions of 93–4; matrix in 20, 91, 96, 97–104, 142; organisation of 94–6
Social Services Organisation Research Unit (SSORU) 93, 94, 96, 104 (n.1), 105 (n.4, 5), 146
Socio-technical systems, design 121, 131, 133
Specialisation 2, 120–1, 213
Specialists 16, 26
Stability 53, 68
Staff utilisation 61, 65, 66–8
Stalker, G. M. 3, 4, 129, 139 (n.3), 173
Stamp, Gillian 146
Star, S. H. 5, 16, 22 (n.6), 110
Steiner, G. A. 22 (n.4), 156 (n.3), 183 (n.2)
Stewart, J. D. 19, 22 (n.10)
Stopford, J. M. 22 (n.7)
Stress 135, 162, 165, 166, 215: and role definition 167, 169 (n.4)
Structural effectiveness, criteria of 113–16, 119–40
Sub-project manager 148
Supervisory group in matrix 30
Supply, concentration of 213, 214
Survey Research Centre (University of Michigan) 139 (n.6)
System(s) 55: closed 129, 130; compatibility with structure 195, 212; design of 121, 198, 200, 207; open 129, 130, 133, 178; reward 51, 55; social 131, 134; support 168, 180–2, 212; technical 131; see also Control system, Information system
Systems theory 93, 118 (n.9), 126, 129

Task forces 3, 144
Tavistock Institute of Human Relations 131, 133, 139 (n.6), 183 (n.5)
Taylor, F. W. 139 (n.1)
Team(s) 132–3: accountability in 86, 87–9; advertising agency account group 76; authority in 87–9; business area 2, 48, 188; child psychiatry 86, 89; collateral 117 (n.2), 144; cross-functional 184, 185; differences in 85; district management 83; egalitarian character 85; integration with organisation 178–9, 185, 190; levels in 89, 134; multi-disciplinary 1, 21, 82–7; in National Health Service 83; power in 89; product 2, 4; programme area 19; project 2, 4, 28, 31, 35, 37, 110, 143; sporting analogy 178; training and development 178–9
Thamhain, H. J. 169 (n.1)
Thom, N. 22 (n.2)
Thompson, J. D. 126, 139 (n.4)
Timmerman, M. 214
Toffler, Alvin 84, 90 (n.3)
Top management, role of 54
Traffic manager 75, 76, 145
Training 178–80, 211: in appraisal interviewing 63; experiential 179; knowledge-based 179; for matrix, example 179–80; organisation of 16
Transfer pricing 203
Trends: economic and social 212; in management 213; trigger incident 46; Trist, Eric 121, 139 (n.6); TRW Systems 132, 147, 157 (n.10), 163; turnover 66; two boss system 124

Uncertainty 3, 54, 71, 90 (n.5), 197: ability to tolerate 167, 171; definition of 127; environmental 127, 129, 136; and power 175; in role relationships 132, 134, 166–7
Unity of command 147
Urwick, Lyndall 139 (n.1)

Vancil, R. F. 203
Vickers, Geoffrey 85
Videlo, Donald 22 (n.5), 33, 150, 152, 157 (n.10)